CONTENTS

Topic 1
Introduction to Geography

CONTENTS

Topic 2
United States and Canada

WORLD GEOGRAPHY
Western Hemisphere

myWorld
INTERACTIVE

Active Journal

SAVVAS
LEARNING COMPANY

ISBN-13: 978-0-328-96497-0
ISBN-10: 0-328-96497-2
11 2021

Topic 3
Middle America

CONTENTS

Topic 4
South America

Topic 5
Europe Through Time

CONTENTS

Topic 6
Europe Today

Topic 7
Northern Eurasia

Introduction to Geography Preview

Essential Question How much does geography affect people's lives?

Before you begin this topic, think about the Essential Question by completing the following activities.

1. What do you think when you see or hear the word *geography*? What does *geography* mean to you? Record your thoughts in the space below.

Map Skills

Using the physical map in the Regional Atlas in your text, label the outline map with the continents and oceans listed. Then color in areas of land and water. Create a key to define what your colors signify.

Africa	Asia
Europe	Australia
North America	South America
Antarctica	Southern Ocean
Pacific Ocean	Atlantic Ocean
Indian Ocean	Arctic Ocean

2. Preview the topic by skimming lesson titles, headings, and graphics. Then place a check mark next to the geographical features that you predict the text will show affecting people's lives. When you finish reading, circle the features that actually do affect people's lives, according to the text.

__citizenship   __climate   __command economy

__culture   __economics   __global positioning system

__government   __language   __maps

__monarchy   __plate tectonics __religion

__time zones   __trade   __water cycle

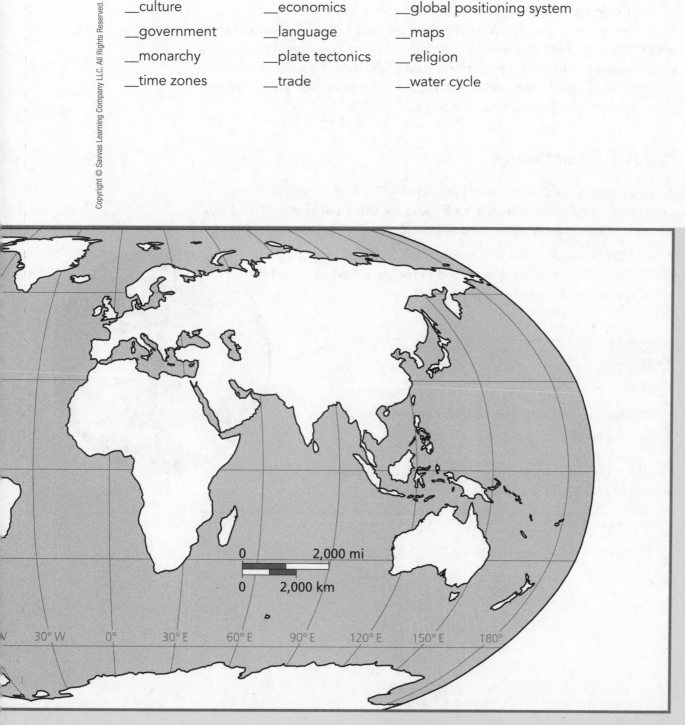

Quest

Project-Based Learning Inquiry

Balancing Development and the Environment

On this Quest, you will work with a team to identify an example of economic development in your area that has an impact on the environment locally or far away. You will ask and research questions about the economic and environmental costs and benefits of the development, and draw conclusions about whether the development is worth the cost. At the end of the Quest, your team will write a blog post in which you present your findings.

1 Ask Questions

As you begin your Quest, keep in mind the Guiding Question: **Can economic development justify its impact on the environment?** and the Essential Question: **How much does geography affect people's lives?**

What other questions do you need to ask in order to draw your conclusion? Two questions are filled in for you. Add at least two more questions for each category.

Theme Benefits of Development

Sample questions:

What positive effects does this development project have on the economy?

How can this development project improve the quality of life for people who live in the area?

Theme Impact on Quality of Life

Theme Harm to the Environment

Theme Financial Impact on the Community

Theme Role of Government

Theme My Additional Questions

 INTERACTIVE

For extra help with Step 1, review the 21st Century Skills Tutorial: **Ask Questions**.

Quest CONNECTIONS

② Investigate

As you read about concepts related to the economy and geography, collect five connections from your text to help you answer the Guiding Question. Three connections are already chosen for you.

Connect to People and the Environment

Lesson 4 Land Use

Here's a connection! Look at the World: Land Use map in your text. Which kinds of land use by people cover large areas, and which kinds of land use cover less area?

How do you think each type of use affects the environment?

Connect to Trade and Development

Lesson 7 Trade Barriers and Free Trade

Here is another connection! Read about economic development and look at the World: Levels of Human Development map. In which areas of the map would you expect to find the greatest human impact on the environment? Why?

What is the environmental impact of expanding trade?

Connect to Government

Lesson 8 Introduction to Government

Read about the powers and responsibilities of governments. What power does a government need to take steps to protect the environment, and what are the limits on those powers?

How can a government balance its responsibility to people's economic well-being with its responsibility to the environment?

It's Your Turn! Find two more connections. Fill in the title of your connections, then answer the questions. Connections may be images, primary sources, maps, or text.

Your Choice | Connect to

Location in text

What is the main idea of this connection?

What does it tell you about balancing development and the environment?

Your Choice | Connect to

Location in text

What is the main idea of this connection?

What does it tell you about balancing development and the environment?

③ Conduct Research

Form teams as directed by your teacher. Meet to decide who will research and write about each theme from Step 1 of the Quest as it relates to your local development project.

You will then research only the theme for which you are responsible. Use the ideas in the connections to explore your theme. Find at least five facts about your development project in relation to your theme. Record key points and reliable sources in the chart below.

Local Economic Development Project:

Theme:

Fact	Source

👆 **INTERACTIVE**

For help with Step 3, review the 21st Century Skills Tutorials: **Search for Information on the Internet** and **Evaluate Web Sites**.

4 Write Your Blog Post

Now it's time to put together all of the information you have gathered and write your blog post.

1. **Prepare to Write** Your team has examined connections and conducted research about the environmental impact of a local economic development project. Together, review your notes and consider your team's position on the development project. Are the benefits worth the environmental costs? Note your thoughts in this space.

Thoughts

2. Write a Draft Write a draft of your blog post. Each team member will draft a paragraph about his or her theme in relation to the development project. Introductory and concluding sentences will express an opinion, and the body of the paragraph will contain supporting facts.

3. Review and Revise Compile the team's paragraphs. Read them and put them in order. Revise them together, correcting any grammatical or spelling errors and adding transition words as needed. As a team, develop introductory and concluding paragraphs that tie it all together relative to the Guiding Question and the Essential Question. Your blog post should present a unified opinion.

4. Create a Visual Now that you have the text for your blog post, find or create a visual to support your key points.

5. Share Finally, publish your blog post, following your teacher's instructions. Read the blog posts by other teams. On a separate sheet of paper, take notes on the information they shared.

6. Reflect Think about your experience completing this topic's Quest. What did you learn about the environmental impact of economic development? What questions do you still have about this subject? How will you answer them?

Reflections

INTERACTIVE

For extra help, review the 21st Century Skills Tutorials: **Work in Teams** and **Publish Your Work**.

Take Notes

Literacy Skills: Summarize Use what you have read to complete the table. Summarize the information for each lesson subheading. The first one has been completed for you.

Geography Basics	
Describing Locations	Geography is the study of human and nonhuman features of Earth. Geographers use the concepts of direction, absolute location, and relative location to describe where locations and objects are.
Geography's Five Themes	
How Do Geographers Show Earth's Surface?	
Understanding Maps	

INTERACTIVE

For extra help, review the 21st Century Skills Tutorial: **Summarize**.

Practice Vocabulary

Use a Word Bank Choose one word from the word bank to fill in each blank. When you have finished, you will have a short summary of important ideas from the section.

Word Bank

cardinal directions	distortion	geography
intermediate directions	latitude	longitude
projection	scale	

North, south, east, and west are the ....

Northwest, northeast, southwest, and southeast are examples of

.... People use them to describe the

location of places. They also use imaginary lines drawn across the

surface of the Earth. Lines that run north to south are called lines of

.... , while those that run east to west are

called lines of ....

People use maps and globes to represent the Earth's surface. Globes

are round like the Earth. They show locations on Earth as they

really are, but at a much smaller ....

To show Earth's round surface on a flat map, a mapmaker must

use a .... , such as the Robinson

or Mercator. Therefore, even the best flat maps still show some

.... of the size or position of objects.

The study of Earth and its human and nonhuman features is called

....

Take Notes

Literacy Skills: Interpret Visual Information Use the diagrams and your text to complete the concept web for Earth's seasons. A concept web for Earth's structure has already been completed. On a separate sheet of paper, make concept webs for forces within Earth that shape it, and forces on Earth's surface that shape it. Add spaces to the concept webs as needed.

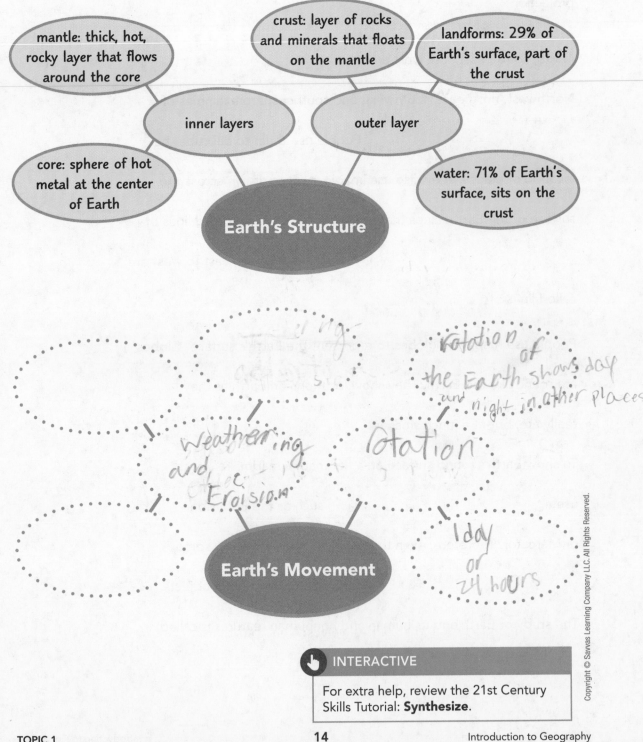

mantle: thick, hot, rocky layer that flows around the core

crust: layer of rocks and minerals that floats on the mantle

landforms: 29% of Earth's surface, part of the crust

inner layers

outer layer

core: sphere of hot metal at the center of Earth

water: 71% of Earth's surface, sits on the crust

Earth's Structure

rotation of the Earth shows day and night in other places

weathering and Erosion

rotation

Earth's Movement

1 day or 24 hours

👆 **INTERACTIVE**

For extra help, review the 21st Century Skills Tutorial: **Synthesize**.

Practice Vocabulary

Sentence Revision Revise each sentence so that the underlined vocabulary term is used correctly. Be sure not to change the vocabulary term. The first one is done for you.

1. <u>Deposition</u> is the process by which water, ice, and wind remove material.

   <u>Deposition</u> is the process by which water, ice, and wind deposit eroded material to create new landforms.

2. When volcanic lava erupts and flows, it is called <u>magma</u>.

   magma ~~lava~~

3. <u>Faults</u> are ~~huge blocks~~ of Earth's crust.

   seams in

4. At the <u>equinox</u>, all parts of Earth are an equal distance from the sun.

   all days/night are equal everywhere

5. <u>Erosion</u> ~~makes rocks break apart into tiny pieces.~~

   removes pieces of rocks.

6. <u>Plate tectonics</u> theory is that Earth's core is made of plates of rock that can move.

   Yes the theory is true.

7. At a <u>solstice</u>, days are nearly equal in length everywhere on Earth.

8. <u>Weathering</u> is a force that carries away soil and bits of rock.

Lesson 3 Climates and Ecosystems

Take Notes

Literacy Skills: Determine Central Ideas Use what you have read to complete the chart. In the box in the second column, identify the central idea of the part of the lesson under the heading in the first column. The first section has been completed for you.

Climate and Weather	Weather is the condition of the air and sky at a certain time and place. Climate is the average weather of a place over many years. Climate graphs show climate for a place over a year's time.
Why Do Temperatures Differ?	Energy from the sun heats earth at different times.
How Does Water Affect Climate?	Warm water can warm the air and cold water can cool they air
Air Circulation and Precipitation	
Types of Climate	There are 3 major climates groups
Biomes and Ecosystems	Ecostsytems change over time (forest, Grassland, cold, climate)

INTERACTIVE

For extra help, review the 21st Century Skills Tutorial: **Identify Main Ideas and Details**.

Practice Vocabulary

Word Map Study the word map for the term *temperate zone*. Characteristics are words or phrases that relate to the term in the center of the word map. Non-characteristics are words and phrases not associated with the term. Use the blank word map to explore the meaning of the term *tropical cyclone*. Then make word maps of your own for these terms: *weather*, *climate*, *tropics*, *water cycle*, *prevailing winds*, *biome*, and *ecosystem*.

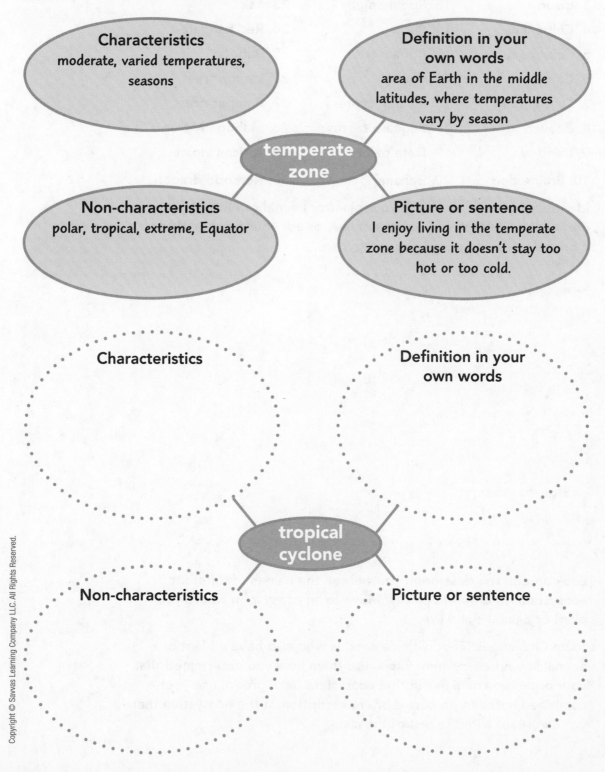

Characteristics
moderate, varied temperatures, seasons

Definition in your own words
area of Earth in the middle latitudes, where temperatures vary by season

temperate zone

Non-characteristics
polar, tropical, extreme, Equator

Picture or sentence
I enjoy living in the temperate zone because it doesn't stay too hot or too cold.

Characteristics

Definition in your own words

tropical cyclone

Non-characteristics

Picture or sentence

Quick Activity Exploring Ecosystems

As directed by your teacher, count off numbers around the classroom, then find the plant or animal below that corresponds to your number.

1. Parrot	11. Redwood	21. Emperor penguin
2. Wildebeest	12. Bearberry	22. Jaguar
3. Bison	13. Red maple	23. Yak
4. Chipmunk	14. Olive	24. Red kangaroo
5. Polar bear	15. Buffalograss	25. Baboon
6. Camel	16. Baobab	26. Lemon tree
7. Orangutan	17. Swamp cypress	27. Pampas grass
8. Beaver	18. Saguaro cactus	28. Mahogany tree
9. Caribou	19. Date palm	29. African violet
10. Prairie dog	20. Banana	30. Komodo dragon

Identify an ecosystem where your plant or animal can live. If you don't know, find reliable sources online, or ask your teacher. Write a brief description of the ecosystem here.

Look around the classroom. You will see the names of different ecosystems in different places. Move to an ecosystem where your plant or animal could live.

Team Challenge! Meet with classmates who also have a plant or animal in your ecosystem. Explain to them how you determined that your organism could live in that ecosystem. As a group, use your combined notes to prepare a brief description of the ecosystem that you can share with the rest of the class.

Take Notes

Literacy Skills: Draw Conclusions Use what you have read to complete the chart. In each column, write details about the topics listed. Then draw conclusions about people's interaction with the environment. The first one has been started for you.

People's Impact on the Environment	Population Growth and Movement
• Humans depend on resources from the environment for survival.	• Their is 7 billon people on earth but we started with less than 10 millon
• We build oil wells, Oil can harm the land and enviroment	• The population grows when their are more people.
• We create pollution that will hurt the land	• before 1950 population was slow a

Conclusions

INTERACTIVE

For extra help, review the 21st Century Skills Tutorial: **Draw Conclusions**.

Practice Vocabulary

Use a Word Bank Choose one term from the word bank to fill in each blank. When you have finished, you will have a short summary of important ideas from the section.

Word Bank

deforestation	urbanization	emigrate
pull factors	natural resources	push factors
biodiversity	industrialization	fossil fuels

.... are useful materials found in the

environment. Nonrenewable resources include minerals, metal ores, and

.... is

the development of machine-powered production and manufacturing.

.... is the loss of forest cover in a region and

can reduce .... , the number of types of living

things in a region or ecosystem. When people leave their home country,

they .... , which means to migrate out of a

place. .... drive people to leave their home

country. .... attract people to new countries.

The movement of people from rural areas to urban areas is called

....

Take Notes

Literacy Skills: Use Evidence Use what you have read to complete the chart. In each column, write details about the elements of culture. Then use this evidence you have gathered from the text to draw a conclusion about what culture means and what makes a culture unique. The first section has been completed for you.

Families and Societies	Language and Religion	Arts, Science, and Technology
Families	Language	Arts
• Families are the basic units of societies.		
• Nuclear families are made up of parents and children.		
• Extended families include grandparents and other relatives.		
Societies	Religion	Science and Technology
"Societies often organize members according to their social class.		

Conclusion

INTERACTIVE

For extra help, review the 21st Century Skills Tutorial: **Support Ideas with Evidence**.

Practice Vocabulary

True or False? Decide whether each statement below is true or false. Circle T or F, and then explain your answer. Be sure to include the underlined vocabulary term in your explanation. The first one is done for you.

> 1. **T / F** A <u>society</u> is the beliefs, customs, practices, and behaviors of a group of people.
> False; a <u>society</u> is a group of humans with a shared culture who have organized themselves to meet their basic needs.

2. **T / F** Members of the same <u>social class</u> may have very different economic conditions.

3. **T / F** Advances in science and technology often raise the <u>standard of living</u>.

4. **T / F** <u>Culture</u> is the basic needs and wants shared by all people, such as food, clothing, and shelter.

5. **T / F** A person's <u>social structure</u> is made of the various social groups the person is part of.

6. **T / F** <u>Cultural diffusion</u> can include the spread of ideas, values, objects, foods, and art forms from one culture to another culture.

Take Notes

Literacy Skills: Compare and Contrast Use what you have read to complete the chart. In each column write the unique characteristics of the economic system listed. Also note any similarities with other economic systems. The first one has been completed for you.

Economic Systems

Traditional Economy	Market Economy	Command Economy	Mixed Economy
• People make economic decisions based on customs and values. • Basic needs are satisfied through traditional methods such as hunting and farming. • People usually do not want to change their way of life, so the economy may be static. • Standard of living is usually low, but there is less waste. • This type of economy is no longer common.	• A market economy is an economy in which individual consumers and Producers make econic decisions • economy market has down sides. • leaving Some people rich ore poor.	• A command Economy is an economy in which the central government makes all economic decisions.	

INTERACTIVE

For extra help, review the 21st Century Skills Tutorial: **Compare and Contrast**.

Practice Vocabulary

Words in Context For each question below, write an answer that shows your understanding of the boldfaced key term.

1. How does **demand** affect the price of a good or service?

2. How is the principle of **opportunity cost** related to economic decision making?

3. What is the role of **consumers** in a pure market economy?

4. What is **supply** in economics?

5. What is the role of **producers** in a country's economy?

6. Why does **economics** have an impact on everyday life?

Quick Activity Your Local Economy

As directed by your teacher, break into small groups to discuss the economy in your town, city, or county. Use the questions below to guide your group discussion.

1. Who are the producers in your local area, and who are the consumers?

2. What goods and services are in high demand in your area?

3. What goods and services are scarce? Describe the prices of these items.

4. Based on what you have read in this lesson, what kind of economy does your local area have? How can you tell?

Team Challenge! As a group, imagine a scenario that would drastically change your local economy. Be creative! What if the entire economic system changed due to a shift in government? What if an essential good or service became scarce? What effects would the change have on the local economy? How might the change affect people's everyday lives or affect the environment? Write a short paragraph or draw a picture that describes your group's imagined economic scenario and the changes it would bring to your local community.

Take Notes

Literacy Skills: Identify Cause and Effect Use what you have read to complete the charts. In each space, write details about the causes that contribute to the listed effect. The first chart has been started for you.

Decision What to Trade and Produce

Scarcity

Scarcity of resources due to limited supply and uneven distribution of resources can make it impossible for a country to produce everything it needs.

Countries must trade the resources, goods, and services that they can produce to obtain resources, goods, and services that they want.

Comparative Advantage

Physical Geography

Increase in Economic Development

Trade Barriers vs. Free Trade

Human Capital Improvements

Resources and Capital Goods

INTERACTIVE

For extra help, review the 21st Century Skills Tutorial: **Analyze Cause and Effect**.

Practice Vocabulary

Word Map Study the word map for the word *tariff*. Characteristics are words or phrases that relate to the word in the center of the word map. Non-characteristics are words and phrases not associated with the word. Use the blank word map to explore the meaning of the word *trade*. Then make word maps of your own for these terms: *comparative advantage, trade barrier, free trade, development, developed country, developing country, gross domestic product* (GDP), and *productivity*.

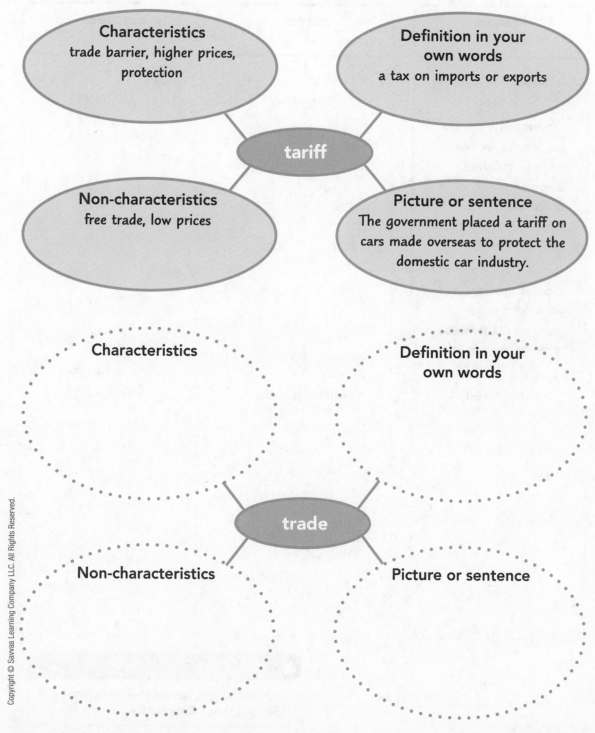

Take Notes

Literacy Skills: Classify and Categorize Use what you have read to complete the chart. In each space write key details that describe the type or characteristic of government listed. The first section has been completed for you.

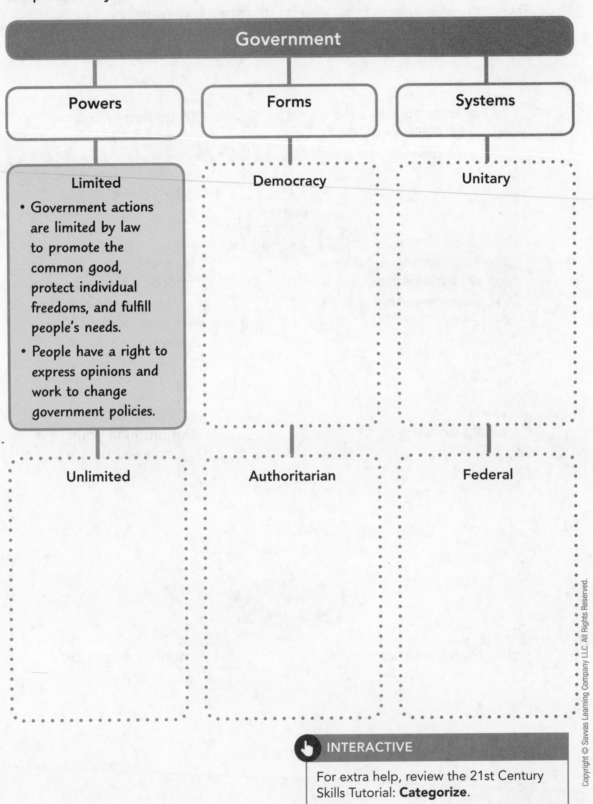

Government

Powers

Limited

- Government actions are limited by law to promote the common good, protect individual freedoms, and fulfill people's needs.
- People have a right to express opinions and work to change government policies.

Unlimited

Forms

Democracy

Authoritarian

Systems

Unitary

Federal

INTERACTIVE

For extra help, review the 21st Century Skills Tutorial: **Categorize**.

Practice Vocabulary

Sentence Revision Revise each sentence so that the underlined vocabulary term is used logically. Be sure not to change the vocabulary term. The first one is done for you.

1. <u>Democracy</u> is an international organization in which citizens hold political power.

   <u>Democracy</u> is a form of government in which citizens hold political power.

2. A <u>constitution</u> identifies the freedoms of individual citizens and outlines the rules and principles citizens must follow.

3. The United Kingdom is an example of <u>authoritarian government</u>, in which all power is held by one person.

4. In a <u>federal system</u>, all power is held by a central government.

5. A <u>monarchy</u> is headed by an elected king or queen.

6. A <u>government</u> is a group of people from the same area who obey the same laws.

7. The <u>unitary system</u> determines whether a government has limited or unlimited power over its citizens.

Take Notes

Literacy Skills: Identify Main Ideas Use what you have read to complete an outline that highlights the main ideas of the lesson. As you create your outline, pay attention to headings, subheadings, and key terms that you can use to organize the information. The first section of the outline has been completed for you.

I. Defining citizenship

    A. A citizen is a legal member of a country.

    B. Citizens of countries have both rights and responsibilities.

II.

INTERACTIVE

For extra help, review the 21st Century Skills Tutorial: **Identify Main Ideas and Details**.

Practice Vocabulary

Matching Logic Using your knowledge of the underlined vocabulary terms, draw a line from each sentence in Column 1 to match it with the sentence in Column 2 to which it logically belongs.

Column 1	Column 2
1. The <u>democratization</u> of societies has been happening gradually for more than 200 years.	Legal members of the United States have several rights and responsibilities.
2. Annika has the full rights and responsibilities of a United States <u>citizen</u>.	You have be at least 18 years old and live in the United States legally for at least ten years to be eligible.
3. Participating in <u>civic life</u> is both a right and a responsibility for citizens of the United States.	Eduardo votes in every election and stays informed about the issues affecting his community.
4. At the <u>naturalization</u> ceremony, 75 adults became U.S. citizens.	After the dictatorship ended, people voted for government representatives who would protect their freedoms.

Take Notes

Literacy Skills: Identify Main Ideas Use what you have read to complete the tables. In each space write one main idea and at least two details that support it. The first one has been completed for you.

Using a Timeline	Organizing Time
Main Idea: Timelines are one method historians use to measure time and determine chronology.	**Main Idea:** The past is divided into prehistory and history.
Details: • Put events in a chronology. • Evaluate patterns and analyze continuity. • Portray a period characterized by specific events.	**Details:** • CE and BCE (common era) (before common era) • different groups have different calendars based on culture.

Historical Sources	Historical Maps	Other Sources
Main Idea: How historians figure out the past	**Main Idea:**	**Main Idea:**
Details: On page 84 in Historical Sources	**Details:**	**Details:**

INTERACTIVE

For extra help, review the 21st Century Skills Tutorial: **Identify Main Ideas and Details**.

Practice Vocabulary

Use a Word Bank Choose one term from the word bank to fill in each blank. When you have finished, you will have a short summary of important ideas from the section.

Word Bank

secondary source	timelines	archaeology
prehistory	period	artifacts
chronology	anthropology	primary sources

.... help historians identify and evaluate

patterns of change. Historians use them to put events in a

..........................., a list of events in the order in

which they occurred. A .... can be

defined by a set of developments that happened during that time.

.... is the time before humans invented

writing and could begin to record events. ....

include letters, diaries, and photographs. ....,

or objects made by human beings, are primary sources. A

.... has information about an event

that does not come from a person who experienced that event.

.... is the scientific study of past

cultures through the examination of artifacts and other evidence.

.... is the study of humankind in all aspects,

especially development and culture.

Writing Workshop Narrative Essay

As you read, build a response to this question: **How have geographic experiences affected your life?** The prompts below will help walk you through the process.

Lessons 1 through 8 Writing Task: Gather Evidence Consider the experiences you have had with elements of geography throughout your life. In the table below, write a sentence or two about each area. You will use one or more of these ideas to write your narrative essay.

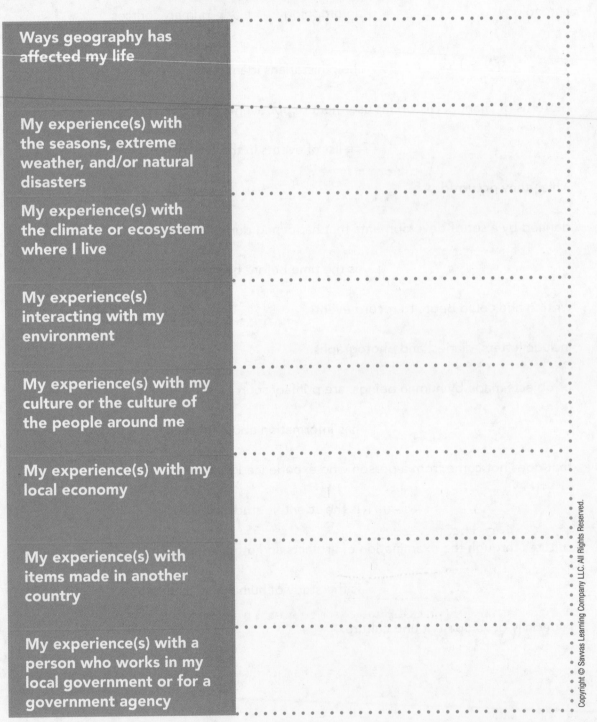

Ways geography has affected my life	
My experience(s) with the seasons, extreme weather, and/or natural disasters	
My experience(s) with the climate or ecosystem where I live	
My experience(s) interacting with my environment	
My experience(s) with my culture or the culture of the people around me	
My experience(s) with my local economy	
My experience(s) with items made in another country	
My experience(s) with a person who works in my local government or for a government agency	

Lesson 9 Writing Task: Use Descriptive Details Descriptive details and sensory language bring life to a story. The reader can "see" and "hear" the things or events the author is describing. Use descriptive details to write about the experience(s) you've had with an element of geography.

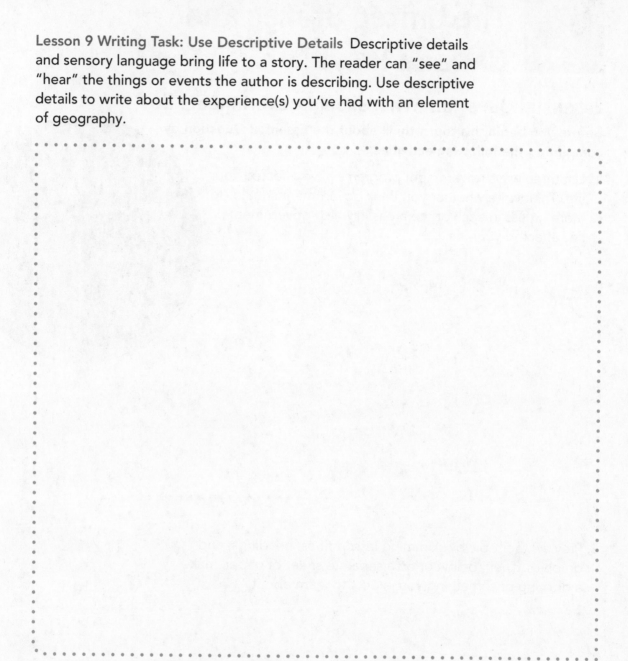

Lesson 10 Writing Task: Add Details Revisit your ideas on the previous page. Add descriptive details to them to make your ideas more complete and interesting.

Writing Task Using your notes from this workshop, write a narrative essay that answers the following question: How have geographic experiences affected your life? As you write, consider using one or more narrative techniques, and describe events using sensory language. The conclusion of your narrative should provide the reader with closure by filling in any missing details needed for the narrative to make sense. The conclusion also sums up the point made by the essay.

The United States and Canada Preview

Essential Question What should governments do?

Before you begin this topic, think about the Essential Question by completing the following activities.

1. List three ways government programs have affected your life. Then write whether you think government should do more or less in each of the areas in which government has affected your life.

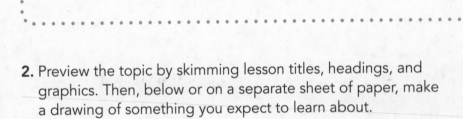

They made me have a close school

2. Preview the topic by skimming lesson titles, headings, and graphics. Then, below or on a separate sheet of paper, make a drawing of something you expect to learn about.

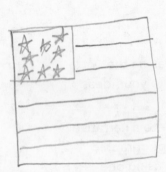

Canada

Map Skills

Using the political and physical maps in the Regional Atlas in your text, label the outline map with the places listed. Then color in the bodies of water and major mountain ranges.

Ottawa   Great Plains   California   Canadian Shield

Rocky Mountains   Alberta   Hudson Bay   Appalachian Mountains

British Columbia   Gulf of Mexico   Bering Strait   Mississippi River

Georgia   Lake Michigan   St. Lawrence River

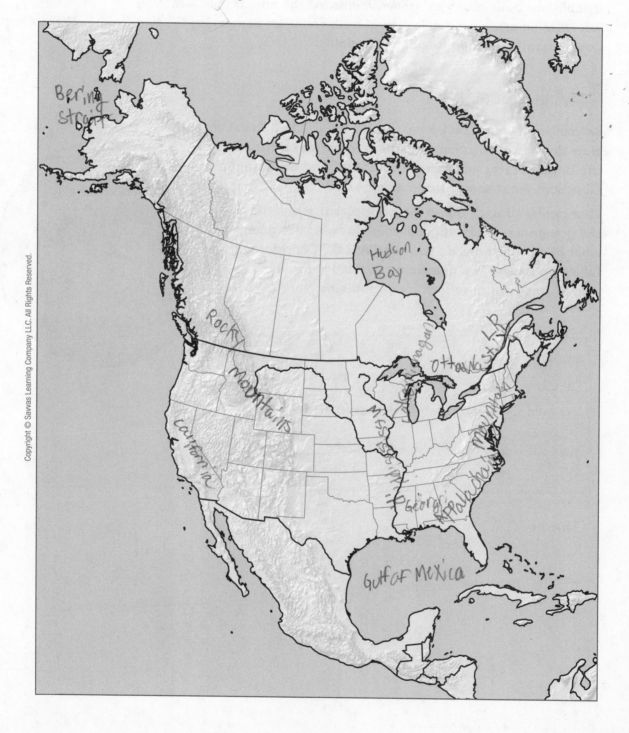

Quest

Document-Based Writing Inquiry

Studying Founding Documents

On this Quest, you will compare the founding documents of the United States and Canada. You will examine as sources documents that established both countries' governments. At the end of the Quest you will write an essay for publication in both countries comparing the two governments and their founding principles.

1 Ask Questions

As you begin your Quest, keep in mind the Guiding Question: **How do the constitutions and other important documents of the United States and Canada compare?** and the Essential Question: **What should governments do?**

Your goal is to understand the principles that shaped the governments of both countries. To meet this goal, what other questions do you need to ask? Consider the following issues. Two questions are filled in for you. Add at least two questions for each theme.

Theme Independence

Sample questions:

Why did the United States seek independence?

How did Canada gain its independence?

Theme Government

what would the canadian goverment do?

what would the American goverment do?

Theme Democracy

How is the candian Democracy different from the usa Democracy?

Theme Rights

What rights do the candians have?

and

what right do the USA people have?

Theme My Additional Questions

 INTERACTIVE

For extra help with Step 1, review the 21st Century Skills Tutorial: **Ask Questions**.

Quest CONNECTIONS

② Investigate

As you read about Canadian and American documents that established new governments, collect five connections from your text to help you answer the Guiding Question. Three connections are already chosen for you.

Connect to Rights

Lesson 4 Canada Unifies

Here's a connection! How did the different ways in which the United States and Canada developed their national identities affect the founding documents of each nation?

① How does expation and conflicts impact peoples
② right
who is respensible for monitering the right
③

② The United States Bill of Rights was ratified in 1791 and certain individual rights were included in Canada's 1982 constitution. What impact do you think ③ the 191-year difference had on the rights Canada listed?

• Why did some America settlen to Mexican area

Connect to Government

Lesson 6 Comparing Systems of Government

Here's another connection! What impact did the founding documents of Canada and the United States have on the structures of the countries' governments?

What's the difference between the way the President is chosen in the United States and the way the prime minister is selected in Canada?

Connect to Social Issues

Lesson 7 Social Challenges

How have the founding documents of the United States and Canada affected each country's ability to deal with its social challenges?

Do the protections provided by the Constitution of the United States apply to people who come to the country illegally?

It's Your Turn! **Find two more connections. Fill in the title of your connections, then answer the questions. Connections may be images, primary sources, maps, or text.**

Your Choice | Connect to

Location in text

What is the main idea of this connection?

What does it tell you about the founding principles of the United States and Canada?

Your Choice | Connect to

Location in text

What is the main idea of this connection?

What does it tell you about the founding principles of the United States and Canada?

③ Examine Primary Sources

Examine the primary and secondary sources provided online or by your teacher. Fill in the chart to show how these sources reflect the origins and founding principles behind the governments of the United States and Canada. The first one is completed for you.

Source	This reflects a founding principle by . . .
The United States Declaration of Independence	saying that *everybody is created equal and is endowed with certain unalienable rights, or rights that cannot be taken away*
A Timeline of the U.S. Constitutional Convention	
The United States Constitution and Bill of Rights	
The Canadian Constitution Act of 1867	
The Canadian Constitution Act of 1982	

👆 INTERACTIVE

For extra help with Step 3, review the 21st Century Skills Tutorial: **Analyze Primary and Secondary Sources**.

4 Write Your Essay

Now it's time to put together all of the information you have gathered and use it to write your essay comparing the founding principles of the United States and Canada.

1. **Prepare to Write** You have collected connections and explored primary and secondary sources about important documents of the United States and Canada and the process by which these documents were developed. Look through your notes and decide which facts you want to highlight in your essay. Record them here.

Facts

2. **Write a Draft** Using evidence from the information in your text and the primary and secondary sources you explored, write a draft of your essay. Be sure to include details about both countries' documents. Include details from the evidence in the material you've studied in this Quest.

3. **Share with a Partner** Exchange your draft with a partner. Tell your partner what you like about his or her draft and suggest any improvements.

4. **Finalize Your Report** Revise your essay based on your partner's comments. Also correct any grammatical or spelling errors.

5. **Reflect on the Quest** Think about your experience completing this topic's Quest. What did you learn about the founding documents of Canada and the United States? What questions do you still have about the two countries and their founding principles? How will you answer them?

Reflections

 INTERACTIVE

For extra help with Step 4, review the 21st Century Skills Tutorial: **Write an Essay**.

Take Notes

Literacy Skills: Classify and Categorize Use what you have read to complete the table. The first one has been completed for you.

Key Features of Culture Regions	
Region	**Key Features**
Arctic	kayaks, hunted sea animals, underground homes, igloos
Subarctic	• hunt moose & elk • witer was cold, summer short
Northwest Coast	• rainy • Platit trees to mak canoes. • They did not fam • totem by there house
California	• lived by coast • eat acorns. lived in hunts. • They moved around • gathering acorns.
Plateau and Great Basin	• lived in Dry mountain. gathered tools. • food was scarce • hunts • hunted Animals.
Southwest	• arid climate • hunt for corn
Great Plains	• live grassblands • l've in village along river. grew corn, squash, tobacco
Northeastern	• warm summer • cold winter • corn, squash beans • deer tukey
Southeastern	• tabacco sunflowel • wood house • clay mud

👆 **INTERACTIVE**

For extra help, review the 21st Century Skills Tutorial: **Categorize**.

Practice Vocabulary

Vocabulary Quiz Show Some quiz shows ask a question and expect the contestant to give the answer. In other shows, the contestant is given an answer and must supply the question. If the blank is in the Question column, write the question that would result in the answer in the Answer column. If the question is supplied, write the answer.

Question

1. What are domed houses made from blocks of snow?

2. _a ceremony in high raking family_

3. _a tent house_

4. What is a home formed by bending the trunks of young trees and tying them together to make a round frame?

5. What type of home is similar to a wigwam but is larger and rectangular?

Answer

1. _Igloos_

2. potlatch

3. tepees

4. _totem Pole_

5. _long house_

Take Notes

Literacy Skills: Synthesize Visual Information Use what you have read and the images you have studied to complete the charts. The first one has been started for you. To complete the first chart analyze the text and the photo on the first page of this lesson. For the second chart, analyze the text and photo on the second page of the lesson.

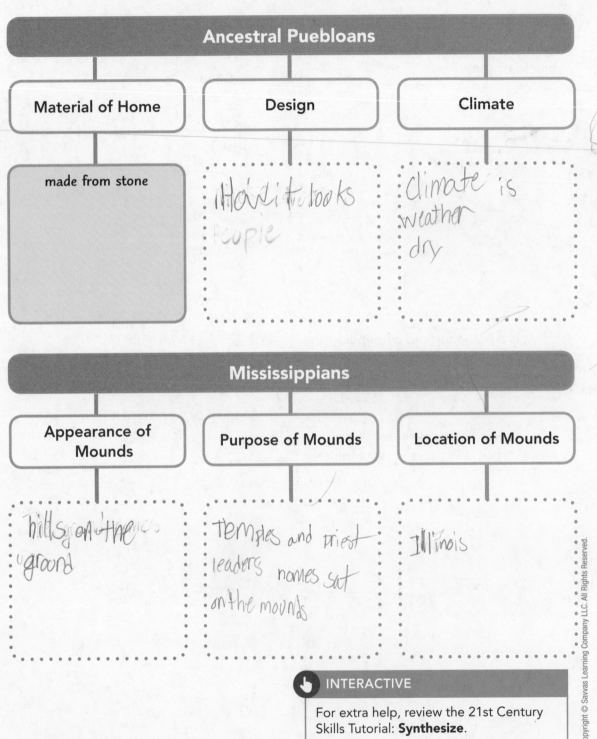

Ancestral Puebloans

Material of Home	Design	Climate
made from stone	How it looks people	climate is weather dry

Mississippians

Appearance of Mounds	Purpose of Mounds	Location of Mounds
hills on the ground	Temples and priest leaders nomes sit on the mounds	Illinois

👆 **INTERACTIVE**

For extra help, review the 21st Century Skills Tutorial: **Synthesize**.

Practice Vocabulary

Matching Logic Using your knowledge of the underlined vocabulary words, draw a line from each sentence in Column 1 to match it with the sentence in Column 2 to which it logically belongs.

Column 1	Column 2
1. Women chose <u>hoyaneh</u> in the Iroquois League.	Women could remove a leader from his position if he did a poor job.
2. Long <u>droughts</u> around the year 1300 made farming difficult for the Ancestral Puebloan people.	They have learned about groups' crops, clothing, tools, and homes.
3. Scientists have learned about North American Indians by studying their <u>artifacts</u>.	This joint council decided important matters.
4. The five Iroquois nations formed the <u>Iroquois League</u> in the 1500s.	Groups left their villages and moved closer to water sources.

Take Notes

Literacy Skills: Synthesize Visual Information Use what you have read and the images you have observed to complete the charts. The first one has been started for you. To complete the first chart analyze the map titled "European Settlement in North America, About 1640" along with the text on that page. For the second chart, analyze the maps of North America in 1753 and 1763 along with the text on that page.

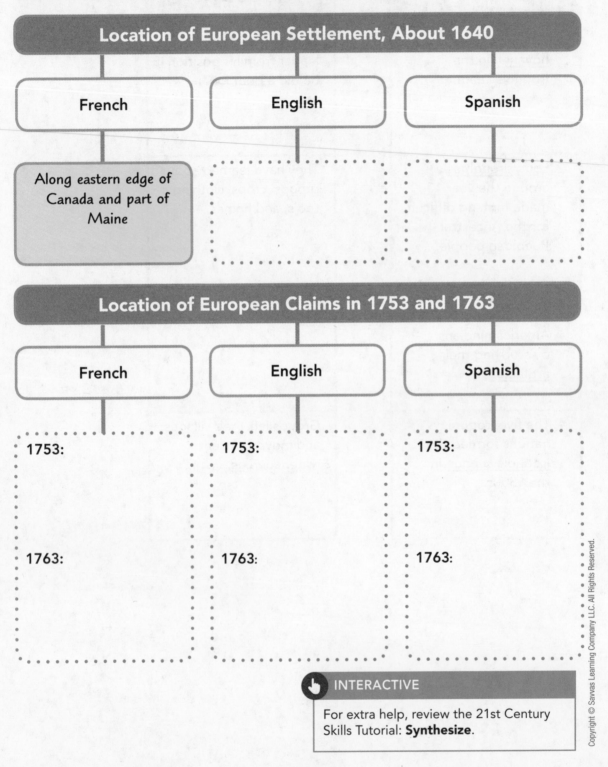

Location of European Settlement, About 1640

French	English	Spanish
Along eastern edge of Canada and part of Maine		

Location of European Claims in 1753 and 1763

French	English	Spanish
1753: 1763:	1753: 1763:	1753: 1763:

👆 **INTERACTIVE**

For extra help, review the 21st Century Skills Tutorial: **Synthesize**.

Practice Vocabulary

True or False? Decide whether each statement below is true or false. Circle T or F, and then explain your answer. Be sure to include the underlined vocabulary term in your explanation. The first one is done for you.

1. **T / F** Eventually, a <u>Northwest Passage</u> was discovered in North America.
False; The French, the Dutch, and the English looked for a <u>Northwest Passage</u>, but eventually it became clear that it did not exist.

2. **T / F** Enslaved Africans were chained together to prevent escapes and <u>mutinies</u>.

3. **T / F** About 500,000 enslaved Africans were transported to the Americas across the <u>Middle Passage</u> between 1500 and 1870.

4. **T / F** <u>Pilgrims</u> settled in Virginia in 1620 and were also called Puritans.

5. **T / F** The <u>triangular trade</u> linked Africa, the Caribbean, and the English colonies.

Quick Activity European Settlement

When European colonists settled in North America, they became residents of a continent that had previously been the home of only American Indians. There were costs and benefits to Europeans settling in North America as well as to American Indians. In the chart below, list both the costs and benefits of European settlement of North America.

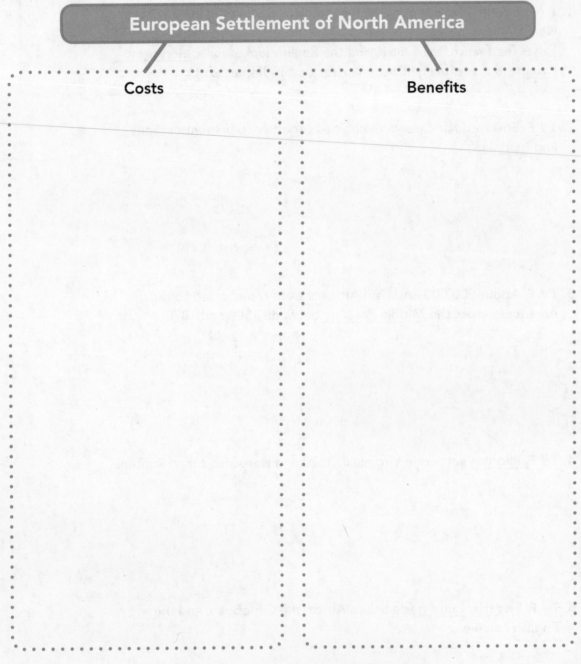

European Settlement of North America

Costs

Benefits

Team Challenge! With a partner, discuss the costs and benefits of European settlement. Be sure to include perspectives from all sides.

August C.

Take Notes

Literacy Skills: Sequence Use what you have read to complete the charts. Enter events in the sequence they occurred. The first one has been started for you.

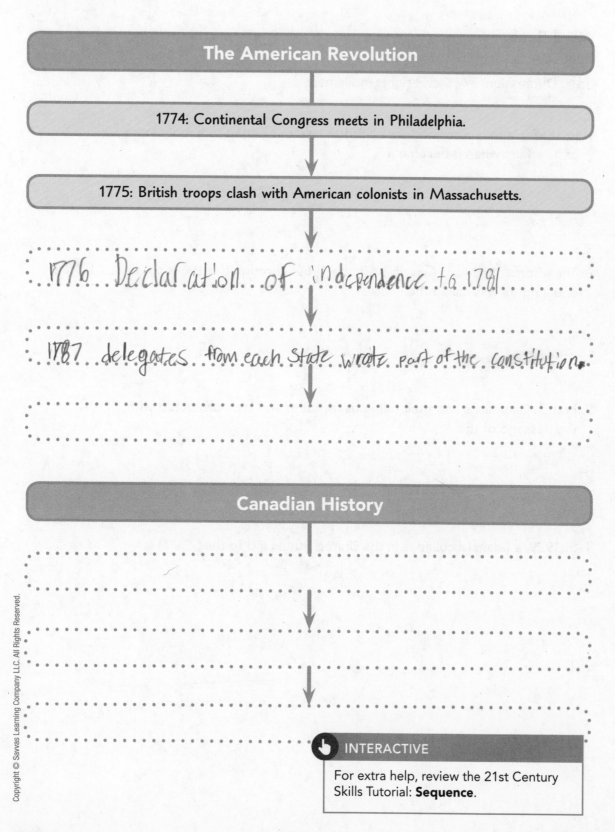

The American Revolution

1774: Continental Congress meets in Philadelphia.

1775: British troops clash with American colonists in Massachusetts.

1776 Declaration of independence to 1781

1787 delegates from each state wrote part of the constitution.

Canadian History

INTERACTIVE

For extra help, review the 21st Century Skills Tutorial: **Sequence**.

Practice Vocabulary

Sentence Builder Finish the sentences below with a key term from this lesson. You may have to change the form of the terms to complete the sentences.

Word Bank

Quebec Act   dominion

Great Depression   civil rights movement

1. In 1867, Canada remained under British rule but began to handle more of its affairs when it became a

2. The efforts of African Americans in the 1950s through the 1960s to win equal rights was called the

3. French Canadians had their religious freedom and laws protected with the passage of the

4. In 1929, a financial collapse in the United States led to the

Take Notes

Literacy Skills: Draw Conclusions Use what you have read to complete the charts. Draw conclusions from the actions in the top boxes. The first one has been started for you.

> People notice good cropland and natural resources in parts of the United States and Canada.

⬇

> People settle in these areas.

⬇

> They tend to get to work on farming and geting natural resources.

> Canada gives priority to immigrants with a college education and knowledge of French or English.

⬇

> They moved to Canada

⬇

>

INTERACTIVE

For extra help, review the 21st Century Skills Tutorial: **Draw Conclusions**.

Practice Vocabulary

Words in Context For each question below, write an answer that shows your understanding of the boldfaced key term.

1. What impact has economic change had on **migration**?

2. Under U.S. **immigration** laws, who is prioritized?

3. What are the characteristics of a **taiga** biome?

Take Notes

Literacy Skills: Compare and Contrast Use what you have read to complete the Venn diagrams with similarities and differences. The first entry has been completed for you.

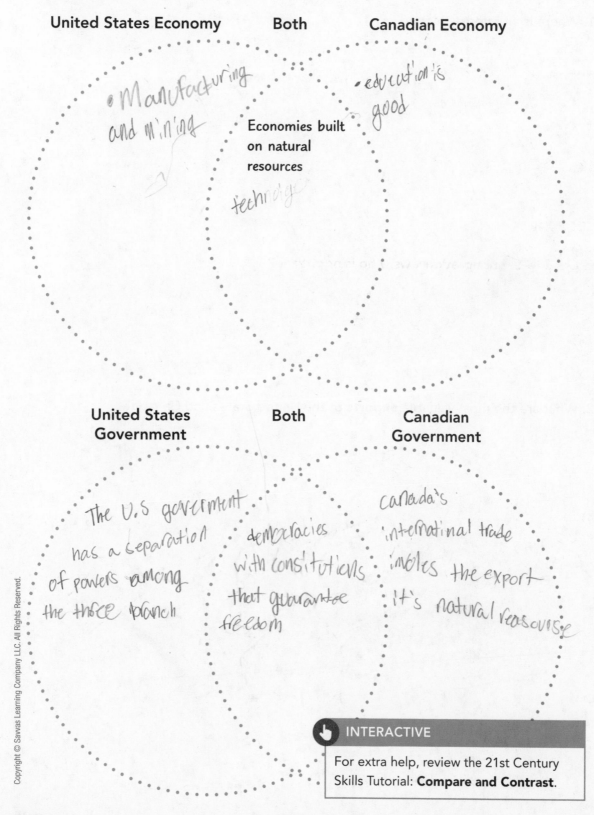

United States Economy   **Both**   **Canadian Economy**

• Manufacturing and mining

Economies built on natural resources

technolgy

• education is good

United States Government   **Both**   **Canadian Government**

The U.S government has a separation of powers among the three branch

democracies with consitutions that guarantee freedom

canada's international trade involes the export it's natural reasourse

INTERACTIVE

For extra help, review the 21st Century Skills Tutorial: **Compare and Contrast**.

Practice Vocabulary

Words in Context For each question below, write an answer that shows your understanding of the boldfaced key term.

1. Where is an **import** shipped?

2. What are the most valuable **exports** of the United States and Canada?

Quick Activity Technology in the Workplace

Over the past 30 years, new technologies have changed the workplace. They've led to the creation of some jobs, but other jobs have been lost. Write a list of some technologies that have developed over the past 30 years.

Team Challenge! Discuss with your classmates the lists everybody came up with. Then, have a conversation about the costs and benefits of these new technologies. Consider the jobs gained in new technological sectors and the jobs lost to automation or software.

Take Notes

Literacy Skills: Use Evidence Use what you have read to complete the charts. Fill in the spaces with evidence that supports the conclusion. The first one has been started for you.

United States and Canada Face Economic Challenges

Taxpayers sometimes resist investing in education.

U.S.A. Has more defense and public safety

the money from investments is not share equally between Americans and canadians.

United States and Canada Face Social Challenges

INTERACTIVE

For extra help, review the 21st Century Skills Tutorial: **Identify Evidence**.

Practice Vocabulary

Vocabulary Quiz Show Some quiz shows ask a question and expect the contestant to give the answer. In other shows, the contestant is given an answer and must supply the question. If the blank is in the Question column, write the question that would result in the answer in the Answer column. If the question is supplied, write the answer.

Question

1. What is the practice of moving a company outside North America, where wages and other costs are lower?

2. What is the money people make through work or investments?

3.

Answer

1.

2.

3. globalization

Writing Workshop Argument

As you read, build a response to this question: **What impact has technology had on the environment in the United States and Canada?** The prompts below will help walk you through the process.

Lessons 1, 2, and 3 Writing Task: Introduce and Support Claims
Write three claims about the impact of technology on the environment in the early history of the United States and Canada, and record supporting evidence. Be sure to use logical reasoning and relevant evidence. You will use these ideas for the argument you will write at the end of the topic.

Claim	Supporting Evidence

Lessons 4 and 5 Writing Task: Distinguish Claims from Opposing Claims and Choose an Organizing Strategy Write a sentence opposing one of your claims. Then write a sentence defending your claim against this opposing claim. On a separate sheet of paper, determine how you will structure your essay to incorporate opposing claims and your responses to them.

Lesson 6 Writing Task: Use Transition Words Write transition words that you could use in your argument to compare and contrast ideas and information or to transition from one argument to another.

Lesson 7 Writing Task: Write a Conclusion Write a paragraph drawing a conclusion about the impact of technology on the environment in the United States and Canada.

Writing Task Building on the work you have done, write an argument of at least four paragraphs on the impact of technology on the environment in the United States and Canada. Cite credible sources to support your arguments.

Essential Question Who should benefit from a country's resources?

Before you begin this topic, think about the Essential Question by completing the following activities.

1. Think about your own community or the country as a whole. Make a list of some of the natural resources that are important. Note who you think benefits most from some of those resources.

Map Skills

Using the political and physical maps in the Regional Atlas in your text, label the outline map with the places listed. Then color in water, desert, and areas of fertile land.

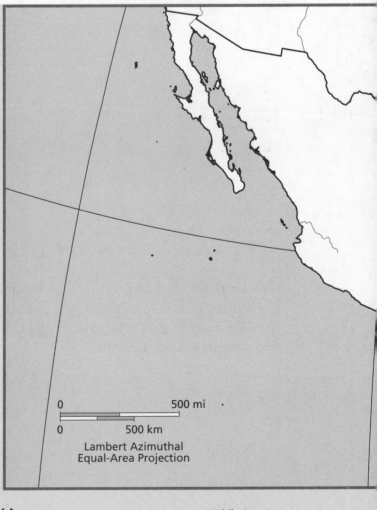

Guatemala

Greater Antilles

Gulf of Mexico

Isthmus of Panama

Dominican Republic

Panama

Yucatán Peninsula

Lesser Antilles

Atlantic Ocean

Costa Rica

Jamaica

Cuba

Caribbean Sea

Pacific Ocean

Mexico

Haiti

0   500 mi

0   500 km

Lambert Azimuthal
Equal-Area Projection

2. Preview the topic by skimming lesson titles, headings, and graphics. Then place a check mark next to the major natural resources you predict the text will cover. After you finish reading the topic, circle the predictions that were correct.

__hydroelectric power   __coal   __oil

__timber   __gas   __soil

__skilled labor   __minerals   __spring water

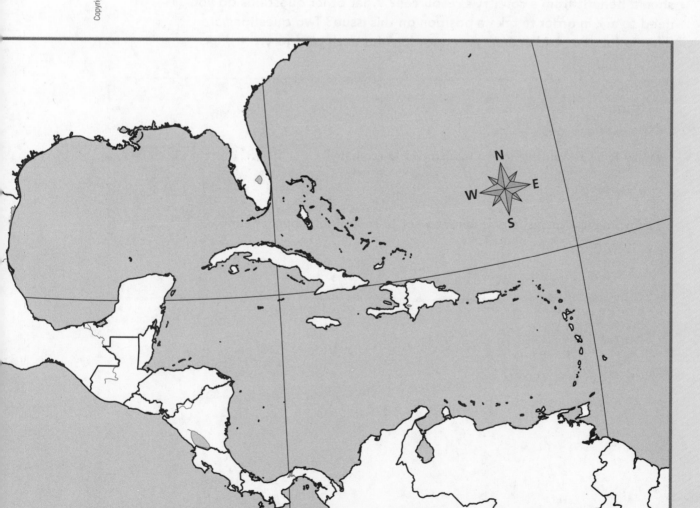

Quest
Discussion Inquiry

Debate Ownership of a Key Resource

On this Quest, you will explore sources and gather information about one of Mexico's key resources. You will take the role of an economist advising the government whether or not to privatize the government-owned oil industry. Then, you will participate in a discussion about the Guiding Question.

1 Ask Questions

As you begin your Quest, keep in mind the Guiding Question: **Should oil production in Mexico be privatized?** and the Essential Question: **Who should benefit from a country's resources?** What other questions do you need to ask in order to take a position on this issue? Two questions are filled in for you. Add at least two questions for each category.

Theme Natural Resources

Sample questions:

Why is oil so valuable in the modern world economy?

Why might a country's government want to control its supply of oil?

Theme Government

Theme Economic Systems

Theme Social Conditions

Theme Wealth

Theme My Additional Questions

 INTERACTIVE

For extra help with Step 1, review the 21st Century Skills Tutorial: **Ask Questions**.

② Investigate

As you read about Middle America, collect five connections from your text to help you answer the Guiding Question. Three connections are already chosen for you.

Connect to the Mexican Revolution

Lesson 4 A Troubled History

Here's a connection! Why do you think the Mexican government took steps to nationalize the oil industry in the years after the revolution?

Who might have favored this decision? Who might have opposed it?

Connect to Cuba's Economy

Lesson 5 The Cuban Revolution

Here's another connection! Why would Castro have opposed outside control of Cuba's industries?

What impact did Castro's decision to nationalize industries have on the Cuban economy?

Connect to Mexico's Economy

Lesson 7 Making a Living in Mexico and Central America

What does this connection tell you about the kind of economy Mexico has? What is the role of oil in Mexico's economy?

What are the benefits and drawbacks of recent changes in Mexico's oil industry?

It's Your Turn Find two more connections. Fill in the title of your connections, then answer the questions. Connections may be images, primary sources, maps, or text.

Your Choice | Connect to

Location in text

What is the main idea of this connection?

What does it tell about what Mexico should consider when debating whether to privatize its oil industry?

Your Choice | Connect to

Location in text

What is the main idea of this connection?

What does it tell about what Mexico should consider when debating whether to privatize its oil industry?

3 Examine Primary Sources

Examine the primary and secondary sources provided online or from your teacher. Fill in the chart to show how these sources provide further information about the costs and benefits of privatizing Mexico's oil industry. The first one has been started for you.

Source	Yes or No? Why?
How Mexico plans to revitalize its energy industry through privatization	YES. Introducing market forces will promote Mexico's gas industry and help the economy grow.
Good News in Mexico	
Mexico's Oil Belongs to Its Citizens, Not the Global 1%	
Privatizing Mexico's Oil Industry Spells Disaster	

INTERACTIVE

For extra help with Step 3, review the 21st Century Skills Tutorial: **Compare Viewpoints**.

4 Discuss!

Now that you have explored sources about the costs and benefits of privatizing Mexico's oil industry, you are ready to discuss with your fellow economists the Guiding Question: **Should oil production in Mexico be privatized?** Follow the steps below, using the spaces provided to prepare for your discussion.

You will work with a partner in a small group of economists. Try to reach consensus, a situation in which everyone is in agreement, on the question. Can you do it?

1. **Prepare Your Arguments** You will be assigned a position on the question, either YES or NO.

My position:

Work with your partner to review your Quest notes from the Quest Connections and Quest Sources.

- If you were assigned YES, agree with your partner on what you think were the strongest arguments from Harrison and *The Washington Times* editors.

- If you were assigned NO, agree on what you think were the strongest arguments from Okón and Buscaglia.

2. **Present Your Position** Those assigned YES will present their arguments and evidence first. As you listen, ask clarifying questions to gain information and understanding.

What is a Clarifying Question?	
These types of questions do not judge the person talking. They are only for the listener to be clear on what he or she is hearing.	
Example: Can you tell me more about that?	Example: You said [x]. Am I getting that right?

 INTERACTIVE

For extra help with Step 4, review the 21st Century Skills Tutorial: **Participate in a Discussion or Debate**.

While the opposite side speaks, take notes on what you hear in the space below.

3. **Switch!** Now NO and YES will switch sides. If you argued YES before, now you will argue NO. Work with your same partner and use your notes. Add any arguments and evidence from the clues and sources. Those *now* arguing YES go first.

When both sides have finished, answer the following:

Before I started this discussion with my fellow economists, my opinion was that oil production in Mexico	*After* I finished this discussion with my fellow economists, my opinion was that oil production in Mexico
_____should be privatized. _____should not be privatized.	_____should be privatized. _____should not be privatized.

4. **Point of View** Do you all agree on the answer to the Guiding Question?

• _____Yes

• _____No

If not, on what points do you all agree?

Take Notes

Literacy Skills: Identify Cause and Effect Use what you have read to complete the charts. For each statement, identify three effects that the cause had on the development of civilizations in Middle America. The first effect has been completed for you.

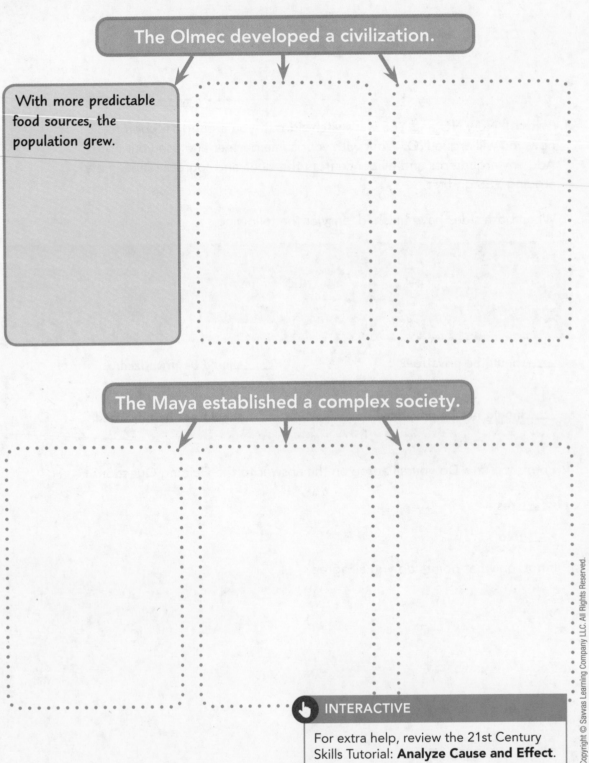

The Olmec developed a civilization.

With more predictable food sources, the population grew.

The Maya established a complex society.

INTERACTIVE

For extra help, review the 21st Century Skills Tutorial: **Analyze Cause and Effect**.

Practice Vocabulary

Matching Logic Using your knowledge of the underlined vocabulary words, draw a line from each sentence in Column 1 to the sentence in Column 2 to which it logically belongs.

Column 1	Column 2
1. The Maya traded across the region for <u>quetzal</u> feathers.	They cleared and set fire to land to enrich the soil for farming.
2. The Maya built temples and <u>observatories</u> in all their cities.	This writing system uses picture representations of words and ideas.
3. The Olmecs used a technique called <u>slash-and-burn agriculture</u>.	They tracked the sun, moon, and stars to know when to plant and harvest.
4. The Maya developed a complex system of <u>hieroglyphics</u>.	These items were important for religious rituals.

Take Notes

Literacy Skills: Summarize Use what you have read to complete the table. List two or three important facts about Aztec civilization for each theme. The first one has been started for you. Then use these notes to write a short summary of the lesson.

Geography	Government and Society	Achievements
• The Aztecs built their capital city in the middle of a lake so they could easily defend it.		

Summary Statement

 INTERACTIVE

For extra help, review the 21st Century Skills Tutorial: **Summarize**.

Practice Vocabulary

Sentence Revision Revise each sentence so that the underlined vocabulary word is used correctly. Be sure not to change the vocabulary word. The first one is done for you.

1. <u>Aqueducts</u> allowed merchants to carry trade over water.

   <u>Aqueducts</u> allowed fresh water to be carried across the salty lake.

2. The Aztecs built their civilization in a <u>basin</u> so that it was easier to do laundry.

3. The Aztecs constructed <u>dikes</u> to keep invaders out of their city.

4. The Aztecs built <u>chinampas</u> to paddle around the lake where they built their city.

Take Notes

Literacy Skills: Sequence Use what you have read to complete the chart. Sequence important events concerning Spanish colonization of Middle America. The first one has been completed for you.

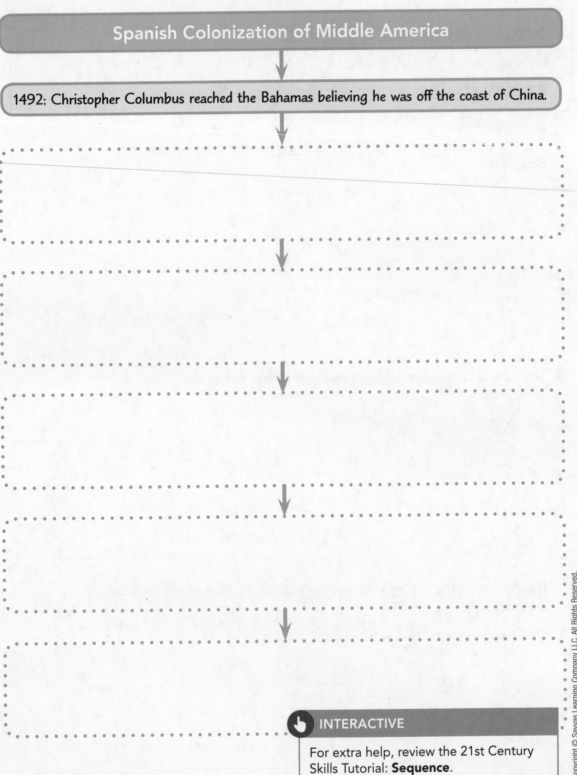

Spanish Colonization of Middle America

1492: Christopher Columbus reached the Bahamas believing he was off the coast of China.

INTERACTIVE

For extra help, review the 21st Century Skills Tutorial: **Sequence**.

Practice Vocabulary

Use a Word Bank Choose one word from the word bank to fill in each blank. When you have finished, you will have a short summary of important ideas from the section.

Word Bank

creoles   Columbian Exchange   conquistadors   mestizos

mulattoes   encomienda   peninsulares

In search of gold, riches, and converts to Christianity, Spanish

.... colonized many territories in Middle America.

They established the .... system, which required

American Indians to pay tribute in goods or labor. Many American Indians

died from disease and harsh treatment. New Spain had a very rigid social

structure. ...., or people born in Spain, were at

the top of the social ladder. Next came ....,

or people born in the Americas to Spanish settlers. The lower social classes

included people of mixed American Indian and European descent who were

called .... and people who were of mixed African

and European descent, who were called ....

Colonization also linked America to the rest of the world, leading to the

............................, in which goods and ideas were exchanged

between the Eastern and Western Hemispheres.

Take Notes

Literacy Skills: Determine Central Ideas Use what you have read to complete the table. Use the supporting details provided to determine the central idea. The first one has been completed for you.

Central Idea	Supporting Details
The Enlightenment inspired revolt and a push toward independence in Mexico and Central America.	• Many people in the Spanish colonies had few rights. • Creoles were tired of being dominated by peninsulares. • People were starting to question the divine right of rulers.
	• Common people in Mexico worked long hours and had very few rights and not enough food. • Juárez wanted to create reforms that would help the poor, but wealthy people were afraid of losing power. • Díaz ignored election results to make someone he favored president.
	• Many newly independent countries looked to the United States to help build their economies. • The United States supported Panama's independence in return for control of a planned canal. • Guatemalan reform and nationalization ended with U.S.-backed overthrow of its elected government.
	• Reforms made in Mexico led to democratic elections in 2000. • There is now free trade in most Central American countries. • Many foreign countries are now investing in Central American countries.

INTERACTIVE

For extra help, review the 21st Century Skills Tutorial: **Identify Main Ideas and Details**.

Practice Vocabulary

Vocabulary Quiz Show Some quiz shows ask a question and expect the contestant to give the answer. In other shows, the contestant is given an answer and must supply the question. If the blank is in the Question column, write the question that would result in the answer in the Answer column. If the question is supplied, write the answer.

Question

1.

2.

3. In 1910, what happened when Díaz tried to install someone he favored as the Mexican president?

Answer

1. nationalizing

2. Mexican Cession

3.

Quick Activity Living Timeline

Based on instructions from your teacher, form a group of three or four students. Use the workspace below to name and briefly explain a few people or events that are important to Mexican history. Each student should choose one to use for this activity. As a group, arrange the events and people in chronological order, or the order in which they happened.

Team Challenge! Construct a living timeline of important people and events in Mexican history. Each team member will be responsible for presenting one person or event. Stand in chronological order and describe to the class your person or event and that person's or event's importance to Mexican history.

Take Notes

Literacy Skills: Text Structure Use what you have read to complete the table. For each heading, write one sentence that describes the importance of that segment. The first one has been completed for you.

How Colonization Changed the Course of Caribbean History	
Segment Heading	**Importance of Events**
Who Were the First Caribbean Peoples?	The first Caribbean peoples were Arawaks and Caribs, who were mostly hunters and gatherers but also planted some staple crops.
European Colonization	
What Was the Transatlantic Slave Trade?	
Larger Countries Win Independence First	
The Cuban Revolution	
Late Moves Toward Independence	

INTERACTIVE

For extra help, review the 21st Century Skills Tutorial: **Take Effective Notes**.

Practice Vocabulary

Word Map Study the word map for the word *dictatorship*.
Characteristics are words or phrases that relate to the word in the
center of the word map. Non-characteristics are words and phrases
that are not associated with the word. Use the blank word map to
explore the meaning of the word *embargo*.

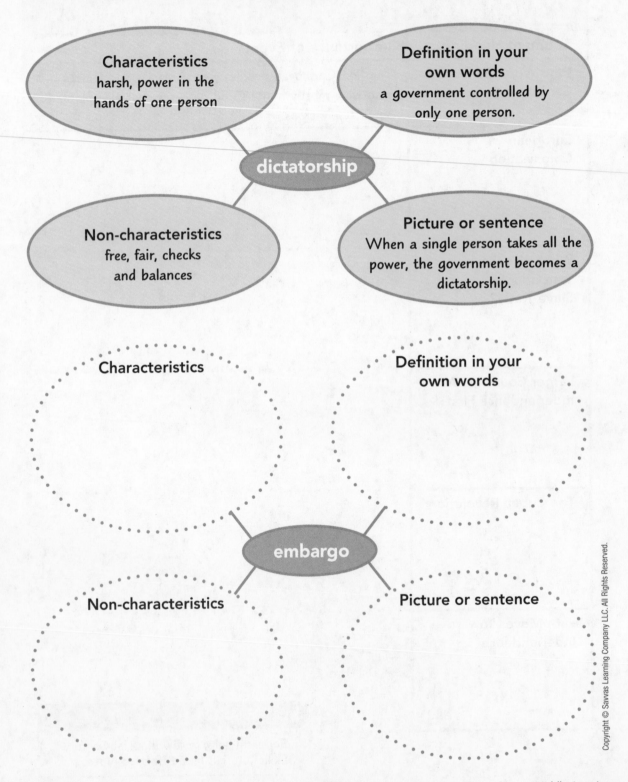

Characteristics
harsh, power in the
hands of one person

Definition in your
own words
a government controlled by
only one person.

dictatorship

Non-characteristics
free, fair, checks
and balances

Picture or sentence
When a single person takes all the
power, the government becomes a
dictatorship.

Characteristics

Definition in your
own words

embargo

Non-characteristics

Picture or sentence

Take Notes

Literacy Skills: Synthesize Visual Information Use what you have read to complete the chart. The chart provides space to take notes on what you learn from each image as you read the text that accompanies it. Notes on the first photo have been started for you. Add any other notes on this photo and notes on the other photo, map, and graph.

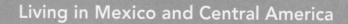

Living in Mexico and Central America

Photos

• The photo shows a densely developed town with narrow streets, so at least some people in this region live in cities.

Map and graphs

INTERACTIVE

For extra help, review the 21st Century Skills Tutorial: **Synthesize**.

Practice Vocabulary

Words in Context For each question below, write an answer that shows your understanding of the boldfaced key term.

1. What is a **diaspora**?

2. What is the effect of **cultural diffusion**?

3. What is a **mural**, and what purpose can it serve?

Quick Activity Quiz Show!

Based on instructions from your teacher, form a group of three or four students. Together you will come up with three questions about the regions of Mexico and Central America to ask the other teams in your class. Make sure you know the answer to your questions.

Team Challenge! **Test your classmates' knowledge. Each team presents their three questions to the rest of the teams, and the other teams try to answer them correctly. After all of the teams have had a chance to present, the team with the most correct answers wins.**

Take Notes

Literacy Skills: Classify and Categorize Use what you have read to complete the charts. For each topic, categorize information about each country listed. The first one has been started for you.

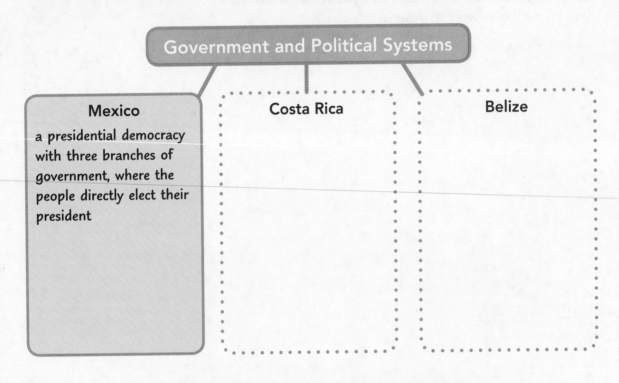

Government and Political Systems

Mexico
a presidential democracy with three branches of government, where the people directly elect their president

Costa Rica

Belize

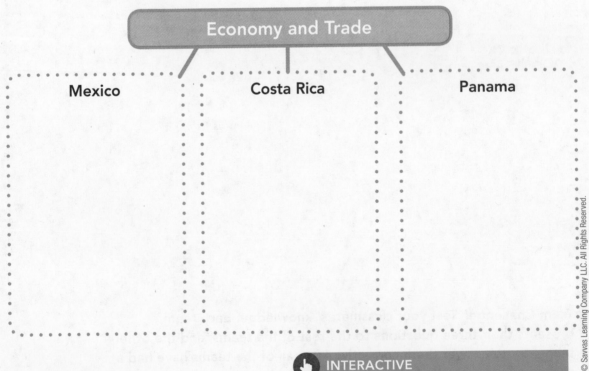

Economy and Trade

Mexico

Costa Rica

Panama

INTERACTIVE

For extra help, review the 21st Century Skills Tutorial: **Categorize**.

Practice Vocabulary

True or False? Decide whether each statement below is true or false. Circle T or F and then explain your answer. Be sure to include the underlined vocabulary word in your explanation. The first one is done for you.

1. **T / F** When someone goes abroad to find work, their family may send them money called <u>remittances</u>.
   False; when someone goes abroad to find work, they may send their family money called <u>remittances</u>.

2. **T / F** <u>Specialization</u> helps with trade because countries focus on producing items they can do well.

3. **T / F** <u>Interdependence</u> for Central American countries means each country produces everything it needs on its own.

4. **T / F** Costa Rica's <u>ecotourism</u> industry is important because it limits the environmental impact of tourism.

Take Notes

Literacy Skills: Use Evidence Use what you have read to complete the table. Supply at least two pieces of evidence from the text to support each claim. The first one has been started for you.

Claim	Evidence
The environment affects economic growth in the Caribbean.	• The area is prone to natural disasters, like hurricanes, that can hurt the economy.
Tourism is a critical part of the economy in Caribbean nations.	
Cuba suffers from a communist government and command economy.	

INTERACTIVE

For extra help, review the 21st Century Skills Tutorial: **Support Ideas with Evidence**.

Practice Vocabulary

Sentence Builder Finish the sentences below with a key term from this section. You may have to change the form of the words to complete the sentences.

Word Bank

Santeria   hurricane   creole

1. A language that mixes elements from other languages is called a

....
.   .
.   .
....

2. One religion that combines Roman Catholic and West African practices is

....
.   .
.   .
....

3. Another word for a tropical cyclone is

....
.   .
.   .
....

Take Notes

Literacy Skills: Draw Conclusions Use what you have read to complete the chart. Identify some of the environmental, economic, and social challenges that the governments and people of Middle America face. Then draw conclusions about the impact of these factors on these countries and the people that live there.

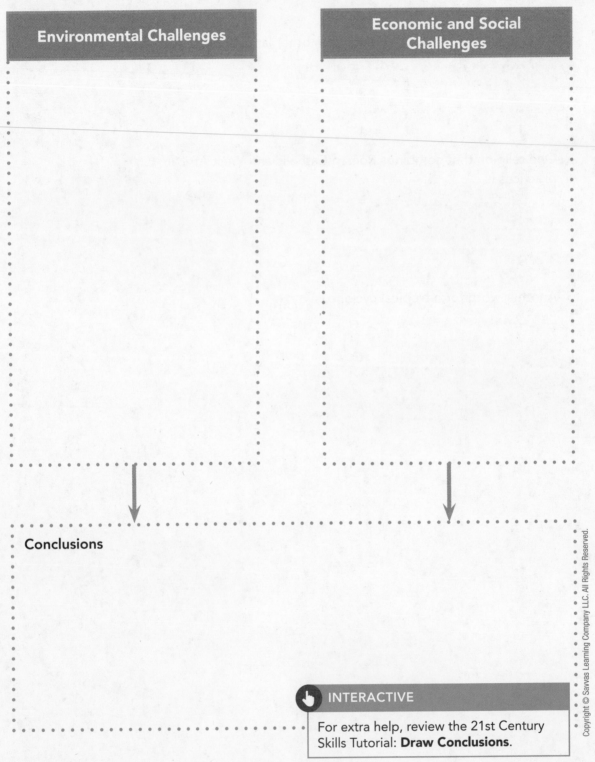

Environmental Challenges

Economic and Social Challenges

Conclusions

INTERACTIVE

For extra help, review the 21st Century Skills Tutorial: **Draw Conclusions**.

Practice Vocabulary

Matching Logic Using your knowledge of the underlined vocabulary words, draw a line from each sentence in Column 1 to match it with the sentence in Column 2 to which it logically belongs.

Column 1	Column 2
1. One of the causes of educational gaps in the region is the <u>digital divide</u>.	Poverty and the illegal drug trade lead to a high rate of crime, which hurts the economy.
2. One effect of earthquakes is that they sometimes cause <u>mudslides</u>.	In some places, fewer than 17 percent of people have access to the Internet at home.
3. <u>Cartels</u> have been responsible for thousands of murders in recent years.	With no tree roots to hold it down, the soil is easily washed away.
4. A consequence of deforestation is <u>erosion</u>.	When lots of dirt slides down the slope of a hill or mountain, homes and whole towns can be buried.
5. A lack of good <u>infrastructure</u> holds the region's economy back.	Reliable transportation, clean water, and electricity attract businesses to a region.

Writing Workshop Explanatory Essay

As you read, build a response to this question: **How does Middle America use its resources?** The prompts below will help walk you through the process.

Lesson 1 Writing Task: Consider Your Purpose Write your thoughts about how people use natural resources. How might you write an explanation of those uses?

Lesson 2 Writing Task: Develop a Clear Thesis Write a clear sentence explaining how people in Middle America have used natural resources. This sentence will be the main point of an essay you will write on this subject at the end of the topic.

Lesson 3 Writing Task: Support Your Thesis with Details Add details about how the Spanish colonists used resources. Review your thesis statement. Revise it if you think that's necessary based on what you've read in this lesson.

Lessons 4 through 7 Writing Task: Pick an Organizing Strategy and Support Your Thesis with Details Use the table to develop an organizing strategy for the body of your essay and to group key details. In the left column, identify three main points to write paragraphs about. Make sure these main points support your thesis. In the right column, note details that support each of these main points. As you work, review your thesis statement and revise it if you think that is necessary.

Lessons 8 and 9 Writing Task: Write an Introduction and a Conclusion Draft your introductory paragraph, which should include your thesis statement and three main points that support it. Draft a conclusion, explaining that resource use is important to Middle America. Use transition words to clarify relationships between ideas.

Writing Task Using the material you've already written, write your explanatory essay on the following topic: How does Middle America use its resources? After you write, read through your essay. Revise it, correcting all spelling and grammar errors.

South America Preview

Essential Question What should governments do?

Before you begin this topic, think about the Essential Question by completing the following activities.

1. Think about your own community and state, as well as the federal government. Make a list of ten things that you know the government does. Choose and rank what you consider to be the five most important things.

2. Preview the topic by skimming lesson titles, headings, and graphics. Then place a check mark next to the government activities that you predict the text will discuss. After reading the topic, circle the predictions that were correct.

__establish military   __pay for art   __abolish slavery

__collect taxes   __establish order   __build monuments

__organize economy   __create colonies   __take control of businesses

Map Skills

Using the political and physical maps in the Regional Atlas in your text, label the outline map with the places listed. Then draw in any major mountain ranges or rivers.

Brazil   Atlantic Ocean   Peru   Buenos Aires

Santiago   Pacific Ocean   Venezuela   Argentina

Caribbean Sea   Colombia   Brasília   The Guianas

KEY

0   800 mi

0   800 km

Lambert Azimuthal
Equal Area Projection

Project-Based Learning Inquiry

Setting Priorities

On this Quest, you are part of a team that will be advising a government committee that is deciding whether or not to host a future Olympics in a South American country. You will gather information about the costs and benefits of hosting the Olympics by examining sources and conducting research. At the end of this Quest, your team will make a recommendation about whether hosting the Olympics is a good economic decision for your country.

① Ask Questions

As you begin your Quest, keep in mind the Guiding Question: **What economic priorities should a government set?** and the Essential Question: **What should governments do?**

What other questions do you need to ask in order to make your recommendation? Two questions are filled in for you. Add at least two questions in each category.

Theme Economic Development

Sample questions:

What factors lead to economic growth in a country?

How can governments best encourage growth?

Theme Culture

Theme Government

Theme Transportation and Urban Development

Theme My Additional Questions

👆 **INTERACTIVE**

For extra help with Step 1, review 21st
Century Skills Tutorial: **Ask Questions**.

② Investigate

As you read about South American countries and the relationships between governments and their economies, collect five connections from your text to help you answer the Guiding Question. Three connections are already chosen for you.

Connect to the Incas

Lesson 2 How Did the Incas Live Together?

Here's a connection! How did the Incan government use the Incas' resources?

Connect to Industrialization

Lesson 4 Dictatorship and Development

Here's another connection! Examine the efforts of dictators to industrialize their countries. How successful were they?

Connect to Venezuela

Lesson 6 South America's Economies

Examine the role Venezuela's government played in the country's economy under Hugo Chávez. What role did the government play in the country's economy under Chávez?

How did supporters and opponents view that role?

It's Your Turn! **Find two more connections. Fill in the title of your connections, then answer the questions. Connections may be images, primary sources, maps, or text.**

Your Choice | Connect to

Location in text

What is the main idea of this connection?

What does it tell you about whether using resources to host the Olympics is a good decision for your country?

Your Choice | Connect to

Location in text

What is the main idea of this connection?

What does it tell you about whether using resources to host the Olympics is a good decision for your country?

③ Conduct Research

Form teams based on your teacher's instructions. Use the table below to develop your research plan. Meet to decide who will research each segment, and list team members' names under their segments. The goal of your research is to help you decide on and support a recommendation on whether or not to host the Olympics.

You will research only the segment that you are responsible for. Be sure to find valid sources and take good notes so you can properly cite your sources. Record key information to help your team make a decision and to include in your written recommendation. Brainstorm ways to enhance your points with visuals.

Segment	Ideas	Sources
Economic Development		
Culture		
Government		
Transportation and Urban Development		

INTERACTIVE

For extra help with Step 3, review the 21st Century Skills Tutorials: **Search for Information on the Internet** and **Make Decisions**.

Quest FINDINGS

4 Write Your Recommendations

Now it's time to put together all of the information you have gathered and write your segment.

1. **Prepare to Write** Review the research you've collected, then meet as a team and decide whether your research supports the government using resources to host the Olympics. Then write notes below on how research on your segment supports the team's decision.

Segment's main point:

Supporting details:

Sources:

Supporting visuals:

2. **Write a Draft** Your segment of the report should fit on one page. That means you will need to get straight to the point. Use transition words and formal grammar.

3. **Share with a Partner** Exchange your segment with a partner. Tell your partner what you like about his or her segment and suggest any improvements. Check for clarity, for evidence-based claims, and for formal language. Revise your segment based on your partner's comments. Correct any grammar or spelling errors and revise your text based on the feedback.

4. **Create a Visual** Now that you have the text of your segment, find or create a visual to support your key points. Aim to make your points more memorable. With your team, decide on a style for the visuals in your committee's report.

5. **Put Together Your Written Recommendation** Once all committee members have written and revised their segments, it's time to put them together. Write an introduction and conclusion paragraph. Ensure smooth transitions from one segment to the next. Review your visuals together.

6. **Reflect on the Quest** After delivering the committee's recommendation, discuss your thoughts. Reflect on the project and list what you might do differently next time so that the project goes more smoothly.

Reflections

INTERACTIVE

For extra help, review the 21st Century Skills Tutorial: **Work in Teams**.

Take Notes

Literacy Skills: Cite Evidence Use what you have read to complete the table. Cite at least three pieces of evidence for each claim about the development of early South American cultures. The first one has been started for you.

Early South American Cultures	
Claim	**Evidence**
Groups of people living in different regions developed different cultures.	• The way people interacted with their environment affected how people lived.
Farming developed in some, but not all, regions of South America.	
Some groups that farmed developed civilizations.	

> 👆 **INTERACTIVE**
>
> For extra help, review the 21st Century Skills Tutorial: **Support Ideas with Evidence**.

Practice Vocabulary

Matching Logic Using your knowledge of the underlined vocabulary words, draw a line from each sentence in Column 1 to match it with the sentence in Column 2 to which it logically belongs.

Column 1	Column 2
1. The <u>Andes</u> are a difficult place to live.	They had no permanent homes and moved around to look for food.
2. Early people began to <u>domesticate</u> wild plants.	The mountains are so high that the elevation makes the air thin and cold.
3. To farm on the mountainsides, people developed <u>terraces</u>.	These flat strips of land allowed people to farm without losing soil and water.
4. Some groups of people continued to be <u>nomads</u>.	Instead of gathering food, they planted seeds and tended gardens.

Take Notes

Literacy Skills: Compare and Contrast Use what you have read to
complete the chart. Write notes about the Incan civilization and
civilizations of the Maya and Aztecs that you read about earlier.
Then, use your notes to write one to three sentences comparing
the civilizations. One note has been completed for you.

Incan Civilization	Maya and Aztec Civilizations
The Incas used a mix of peaceful and military means to build a huge empire.	

Compare and Contrast

 INTERACTIVE

For extra help, review the 21st Century
Skills Tutorial: **Compare and Contrast**.

Practice Vocabulary

True or False? Decide whether each statement below is true or false. Circle T or F, and then explain your answer. Be sure to include the underlined vocabulary word in your explanation. The first one is done for you.

1. **T / F** A quipu was how the Incas kept records.
   True; a quipu was a record-keeping device, made of knotted strings, used to track goods and people.

2. **T / F** A hierarchy is a way to carry water up the mountainside.

3. **T / F** An ayllu is the name of an animal similar to a llama or alpaca.

4. **T / F** The Inca paid taxes through the mita system.

Take Notes

Literacy Skills: Identify Cause and Effect Use what you have read to complete the chart. List five specific results of European colonization of South America and the effects they had on South American Indians. The first one has been completed for you.

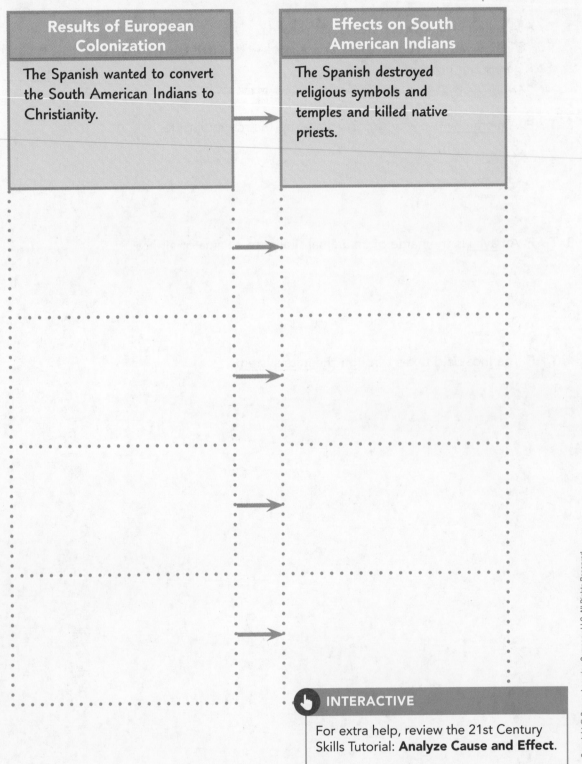

Results of European Colonization	Effects on South American Indians
The Spanish wanted to convert the South American Indians to Christianity.	The Spanish destroyed religious symbols and temples and killed native priests.

INTERACTIVE

For extra help, review the 21st Century Skills Tutorial: **Analyze Cause and Effect**.

Practice Vocabulary

Sentence Revision Revise each sentence so that the underlined vocabulary word is used correctly. Be sure not to change the vocabulary word. The first one is done for you.

1. Millions of South American Indians survived smallpox because they had <u>immunity</u>.
Millions of South American Indians *died from* smallpox because they lacked <u>immunity</u>.

2. The <u>Treaty of Tordesillas</u> was a treaty the Spanish signed with the Incas.

3. <u>Bandeiras</u> were Portuguese explorers who were aided by native peoples.

4. The <u>Line of Demarcation</u> divided Africa in half.

5. Early settlers built houses out of <u>brazilwood</u>.

Quick Activity Sort It Out!

With your classmates, sort out which South American countries that you read about in this lesson were colonized by which European country, based on information provided in this lesson and the South America: Political map in the Regional Atlas.

Your teacher will assign students to represent French Guiana, Suriname, Guyana, Venezuela, Colombia, Ecuador, Peru, Bolivia, Chile, Argentina, Paraguay, Brazil, and Uruguay; and one student each to the colonial empires: Spain, Portugal, the Netherlands, France, and Britain. Make a sign with your country's name in big, bold letters so it can be read from across the room. Students representing the colonial empires should stand in five different parts of the room. Students representing the present or former colonies should move to the correct empire.

Team Challenge! Divide evenly into two teams. Which side can remember the greatest number of impacts of colonization on South America? Your teacher or a student will record the list and keep score.

Take Notes

Literacy Skills: Sequence Use what you have read to complete the chart. Sequence the important events leading up to the independence of the South American colonies. The first one has been completed for you.

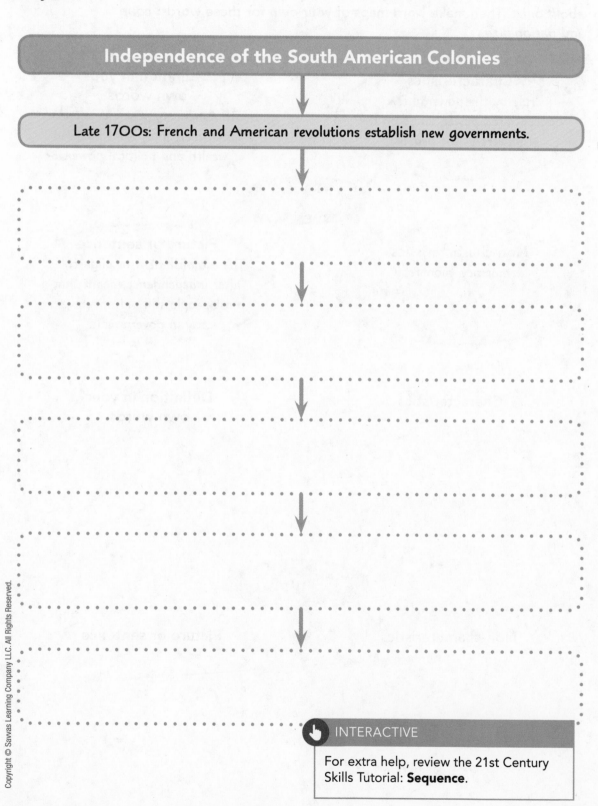

Independence of the South American Colonies

↓

Late 1700s: French and American revolutions establish new governments.

↓

↓

↓

↓

👆 INTERACTIVE

For extra help, review the 21st Century Skills Tutorial: **Sequence**.

Practice Vocabulary

Word Map Study the word map for the word *oligarchy*. Characteristics are words or phrases that relate to the word in the center of the word map. Non-characteristics are words and phrases not associated with the word. Use the blank word map to explore the meaning of the word *abolitionist*. Then make word maps of your own for these words: *coup* and *nationalize*.

Characteristics
rule by the powerful few, concentrated power, limits on the powers of the people

Definition in your own words
South American, post-independence, landowner group that hoarded wealth and political power

oligarchy

Non-characteristics
democracy, monarchy

Picture or sentence
The establishment of oligarchies after independence meant that most people still did not have a say in government.

Characteristics

Definition in your own words

abolitionist

Non-characteristics

Picture or sentence

Take Notes

Literacy Skills: Classify and Categorize Use what you have read to complete the chart. List at least three facts about each region that distinguish it from other areas of South America. Be sure to include information about languages spoken and population distribution. The first one has been started for you.

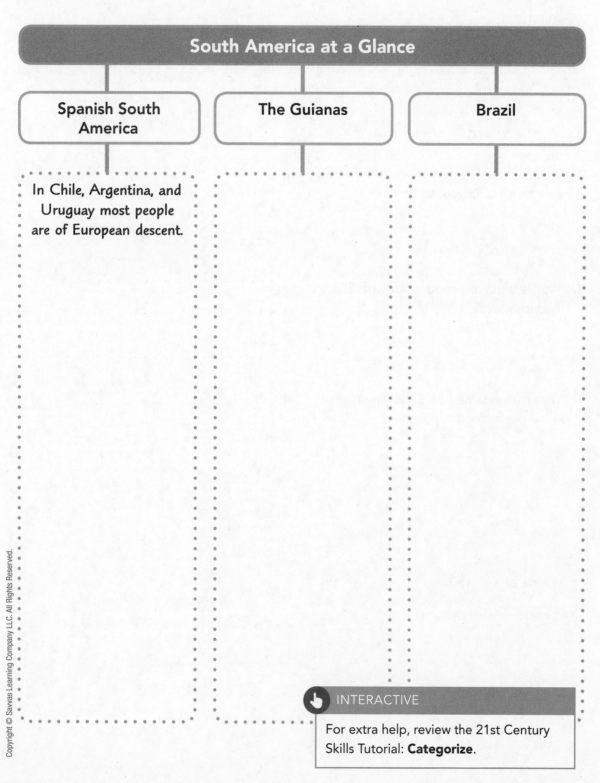

South America at a Glance

Spanish South America

The Guianas

Brazil

In Chile, Argentina, and Uruguay most people are of European descent.

INTERACTIVE

For extra help, review the 21st Century Skills Tutorial: **Categorize**.

Practice Vocabulary

Vocabulary Quiz Show Some quiz shows ask a question and expect the contestant to give the answer. In other shows, the contestant is given an answer and must supply the question. If the blank is in the Question column, write the question that would result in the answer in the Answer column. If the question is supplied, write the answer.

Question	Answer
1.	1. Carnival
2. What is it called when machines do the work that humans used to do?	2.
3. Brazil is a major producer of what kind of soap opera?	3.
4. What is the name of a Brazilian style of dance performed in parades?	4.
5.	5. ethnic group

Take Notes

Literacy Skills: Determine Central Ideas Use what you have read to complete the charts. Using complete sentences, identify three central ideas from the text that support each headline. The first one has been started for you.

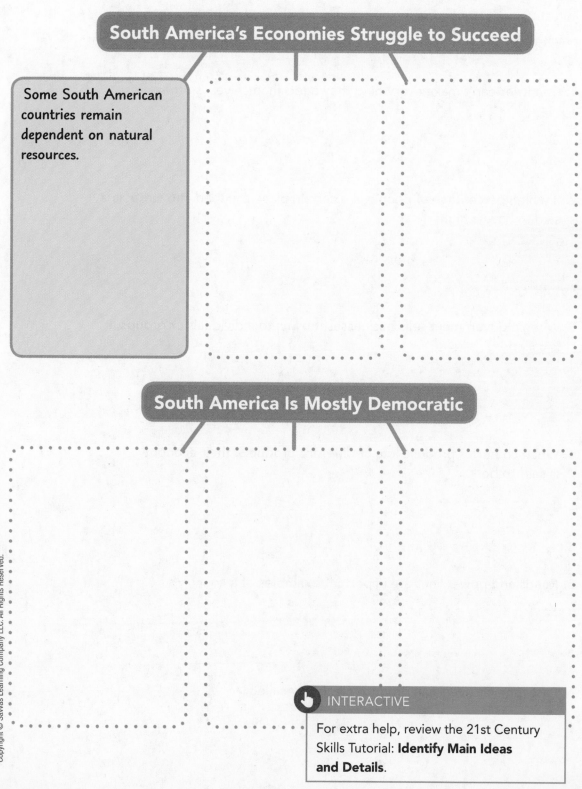

South America's Economies Struggle to Succeed

Some South American countries remain dependent on natural resources.

South America Is Mostly Democratic

INTERACTIVE

For extra help, review the 21st Century Skills Tutorial: **Identify Main Ideas and Details**.

Practice Vocabulary

Sentence Builder Finish the sentences below with a key term from this section. You may have to change the form of the words to complete the sentences.

Word Bank

infrastructure   hydroelectric power   literacy rate

privatizing   interdependent   diversified

1. Countries can't make everything they need themselves, so they are

2. If a high percentage of people in a certain place can read and write, it is said to have a high

3. When a government sells businesses it owns to individuals or groups, it is called

4. When countries produce lots of different products, their economy is said to be

5. Roads and power lines are important examples of a country's

6. Fast-running rivers can produce electricity called

Quick Activity Workers Needed!

Examine the map showing the land use and major resources
of South America.

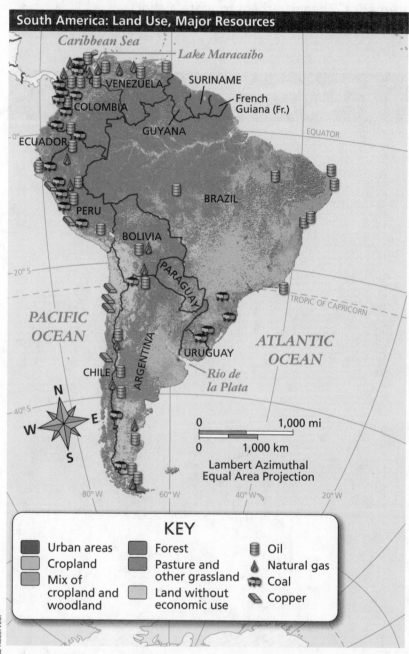

South America: Land Use, Major Resources

Caribbean Sea
Lake Maracaibo
VENEZUELA
SURINAME
French
Guiana (Fr.)
COLOMBIA
GUYANA
EQUATOR
0°
ECUADOR
PERU
BRAZIL
BOLIVIA
PACIFIC
OCEAN
20° S
PARAGUAY
TROPIC OF CAPRICORN
ARGENTINA
URUGUAY
ATLANTIC
OCEAN
CHILE
Rio de
la Plata
N
W   E
40° S
S
0   1,000 mi
0   1,000 km
Lambert Azimuthal
Equal Area Projection
80° W   60° W   40° W   20° W

KEY

Urban areas	Forest
Cropland	Pasture and other grassland
Mix of cropland and woodland	Land without economic use

🛢 Oil
💧 Natural gas
Coal
Copper

Team Challenge! With a partner, look at this map and think of
a business you could start by taking advantage of one of South
America's resources. Write a short ad to post on a business website
to attract investors to invest money in your business. Mention where
your business will be located and the resource it will use.

Take Notes

Literacy Skills: Summarize Use what you have read to complete the chart. Write two or three sentences that summarize each category of challenges facing South America. Then use them to write a short summary of the lesson. The first sentence has been completed for you.

Political Challenges	Environmental Challenges	Economic Challenges
Corruption is a widespread problem in many South American countries.		

Summary

INTERACTIVE

For extra help, review the 21st Century Skills Tutorial: **Summarize**.

Practice Vocabulary

Words in Context For each question below, write an answer that shows your understanding of the boldfaced key term.

1. How has **corruption** affected countries in South America?

2. Why are **nongovernmental organizations** important?

3. What are the causes of **deforestation**?

4. Why are **free trade zones** important?

Writing Workshop Argument

As you read, build a response to this question: **Should a government have a strong or weak role in its country's economy?** The prompts below will help walk you through the process.

Lessons 1 and 2 Writing Task: Introduce Claims Think about how governments affect the economy. How strong a role do you think they should play in the economy? List your ideas below.

Lesson 3 Writing Task: Develop a Clear Thesis Consider what you have learned about government involvement in South America's economies. Write a thesis statement—a clear sentence stating your main argument—on the role you think a government should play in an economy.

Lesson 4 Writing Task: Distinguish Claims From Opposing Claims Consider the strong and weak government roles described in this lesson. On a separate sheet of paper, write at least one benefit and one drawback of each kind of system.

Lessons 5 through 7 Writing Task: Support Claims List details from your reading, including from earlier lessons, that support the claim you made in your thesis statement.

Writing Task Think about what you have learned about the economic roles of different South American governments. Using your thesis and the supports for your claim, answer the following question in a five-paragraph argument on a separate sheet of paper: Should a government have a strong or weak role in its country's economy?

Europe Through Time Preview

Essential Question How should we handle conflict?

Before you begin this topic, think about the Essential Question by completing the following activities.

1. Consider the times you've experienced a conflict or a disagreement with someone in your life. Think about three of those times and write two or three sentences about each of them. Circle the example in which you were most satisfied with the way you handled the situation.

2. Preview the topic by skimming lesson titles, headings, and graphics. Then, place a check mark next to the examples that you predict will show ways that conflicts have been handled. After you finish reading the topic, circle the predictions that were correct.

__The Greek city-states of Sparta and Athens had different governments and values.

__The Romans built aqueducts to bring water to their cities.

__In Europe, the western and eastern Christian churches split apart.

__The king of England signed the English Bill of Rights.

__The Peace of Augsburg divided Germany into Catholic and Protestant states.

__People in Eastern Europe protested against communist rule.

__Countries formed the European Union.

Map Skills

Using the political and physical maps in the Regional Atlas in your text, label the outline map with the places listed. Then, color in water and areas of fertile land.

Alps

Black Sea

English Channel

Great Britain

France

Rome

Paris

Iberian Peninsula

Balkan Peninsula

Germany

North European Plain

Scandinavian Peninsula

Mediterranean Sea

Ireland

Danube River

Baltic Sea

Italian Peninsula

Poland

Greece

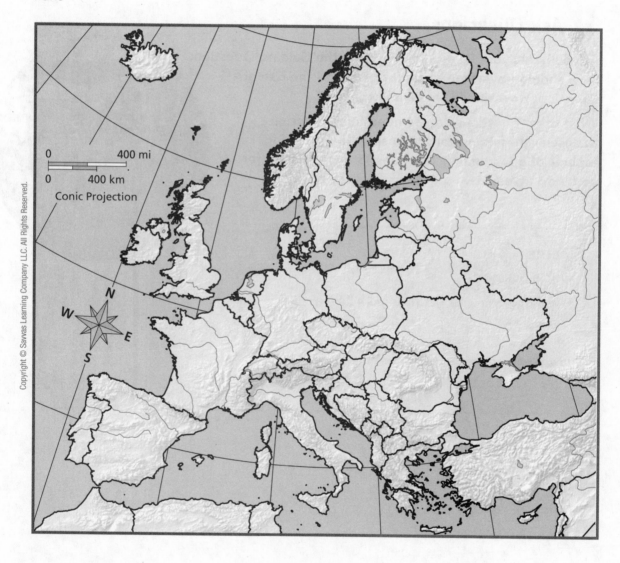

Quest
Document-Based Writing Inquiry

Planning a New Government

On this Quest, you will explore different forms of government. You will examine sources about different forms of government. By the end of the Quest, you will determine which form of government you believe works best and write an official recommendation on the form of government you think a newly independent country should adopt.

1 Ask Questions

As you begin your Quest, keep in mind the Guiding Question: **How should governments be formed?** and the Essential Question: **How should we handle conflict?**

What other questions do you need to ask in order to answer these questions? Consider the following aspects of government. Two questions are filled in for you. Add at least two questions for each category.

Theme Effectiveness

Sample questions:

How do governments get their work done?

How do governments keep order?

Theme Citizens' Rights

Theme Security

Theme Economic Well-Being

Theme The Common Good

Theme My Additional Questions

 INTERACTIVE

For extra help with Step 1, review
the 21st Century Skills Tutorial:
Ask Questions.

2 Investigate

As you read about forms of government, collect five connections from your text to help you answer the Guiding Question. Three connections are already chosen for you.

Connect to Plato's Opinion on Leadership

Lesson 2 Greek Culture and Achievements

Here's a connection! What qualities did Plato believe a good leader should have?

Why do you think these qualities are important in government?

Connect to English Bill of Rights

Lesson 6 Powerful Kingdoms

Here's another connection! How did the government of England in the late 1600s differ from governments in the rest of Europe at the time?

What would be the advantages and disadvantages of a constitutional monarchy?

Connect to the Enlightenment

Lesson 7 Science and the Enlightenment

What did Enlightenment thinkers say about government?

What effect did Enlightenment thinkers have on the establishment of governments in North America and elsewhere?

It's Your Turn! **Find two more connections. Fill in the title of your connections, then answer the questions. Connections may be images, primary sources, maps, or text.**

Your Choice | Connect to

Location in text

What is the main idea of this connection?

What does it tell you about forms of government?

Your Choice | Connect to

Location in text

What is the main idea of this connection?

What does it tell you about forms of government?

3 Examine Primary Sources

Examine the primary and secondary sources provided online or from your teacher. Fill in the chart to show how these sources provide further information about forms of government. The first one has been started for you.

Source	The form of government supported and the reasons for that support:
Funeral Oration	democracy because it provides justice to all and opens government office to anyone with abilities
The Republic	
Leviathan	
Two Treatises of Government	
Origins of Totalitarianism	

INTERACTIVE

For extra help with Step 3, review the 21st Century Skills Tutorial: **Analyze Primary and Secondary Sources**.

4 Write Your Essay

Now it's time to put together all of the information you have gathered and use it to write your report.

1. **Prepare to Write** You have collected connections and explored primary and secondary sources about different forms of government. Look through your notes and decide which ideas you want to highlight in your report and what kind of government you are going to recommend. Record them here.

Ideas:

Recommendation:

2. Write a Draft Using evidence from the textbook and the primary and secondary sources you explored, write a draft of your report. Be sure to discuss various forms of government, not just your recommendation. Include details from the evidence in the material you've studied in this Quest.

3. Share with a Partner Exchange your draft with a partner. Tell your partner what you like about his or her draft and suggest any improvements.

4. Finalize Your Report Revise your essay based on the comments you receive from your partner. Correct any grammatical or spelling errors.

5. Reflect on the Quest Think about your experience completing this topic's Quest. What did you learn about the different forms of government? What questions do you still have about forms of government? How will you answer them?

Reflections

 INTERACTIVE

For extra help with Step 4, review the 21st Century Skills Tutorial: **Write an Essay**.

Take Notes

Literacy Skills: Main Ideas and Details Use what you have read to complete the table. In each row write details that support the main idea provided. The first one has been completed for you.

Main Idea	Details
The first people arrived in Europe about 1 million years ago, and over time, societies developed.	• The first humans to arrive in Europe came from Africa by way of Asia. • Over time, these early humans developed into Neanderthals. • Eventually, *Homo sapiens* came out of Africa and displaced Neanderthals. • *Homo sapiens*, who had larger brains, learned how to use fire, made cave paintings, and held religious beliefs.
Farming changed Europe.	
Europeans produced better and stronger tools.	

👆 **INTERACTIVE**

For extra help, review the 21st Century Skills Tutorial: **Identify Main Ideas and Details**.

Practice Vocabulary

Words in Context For each question below, write an answer that shows your understanding of the boldfaced key term.

1. Who were the **Neanderthals**?

2. Where in Europe did *Homo sapiens* live?

3. What happened in Europe during the **Bronze Age**?

Take Notes

Literacy Skills: Determine Central Ideas Use what you have read to complete the charts. In the top box, write the central idea. Then, complete the lower boxes with missing details. Both charts have been started for you.

Solon ended the practice of selling into slavery poor people who could not pay their debts.

Cleisthenes increased the number of citizens who could vote.

The Greeks created a mythology to explain the world.

Philosophers such as Socrates, Plato, and Aristotle used reason, or logic, to understand reality.

INTERACTIVE

For extra help, review the 21st Century Skills Tutorial: **Determine Central Ideas**.

Practice Vocabulary

Vocabulary Quiz Show Some quiz shows ask a question and expect the contestant to give the answer. In other shows, the contestant is given an answer and must supply the question. If the blank is in the Question column, write the question that would result in the answer in the Answer column. If the question is supplied, write the answer.

Question	Answer
1. What is the term for an independent state consisting of a city and its surrounding territory?	1.
2. What is the worship of many gods?	2.
3.	3. aristocracy
4.	4. mythology
5. What is membership in a community that gives a person civil and political rights and obligations?	5.
6. What is a system of rule by the people?	6.
7.	7. Socratic method

Take Notes

Literacy Skills: Summarize Use what you have read to complete the tables. Note the most important details from the section provided. The first row has been completed for you.

How Was the Republic Governed?	Important Details
Rome's Constitution	• Government included strong leaders, aristocrats, and average people. • Government was organized by basic rules and principles. • Constitution was based on tradition and custom, not written.
Principles of Republican Government	
Citizens and Officials	

The Empire and the Roman Peace	Important Details
Augustus and the Pax Romana	
Other Emperors	

👆 **INTERACTIVE**

For extra help, review the 21st Century Skills Tutorial: **Summarize**.

Practice Vocabulary

True or False? Decide whether each statement below is true or false. Circle T or F, and then explain your answer. Be sure to include the underlined vocabulary word in your explanation. The first one is done for you.

1. **T / F** A <u>republic</u> is a government in which a monarch, such as a king, makes every decision.

   False; a republic is a government in which citizens have a right to vote and elect officials.

2. **T / F** An <u>empire</u> is a state that contains only one country.

3. **T / F** A <u>constitution</u> is a system of basic rules and principles by which a government is organized.

Take Notes

Literacy Skills: Draw Conclusions Use what you have read to complete the charts. The top box lists a conclusion you might draw from the text. In the bottom box, list evidence from the text that supports each conclusion. The first one has been started for you.

> **Two separate Christian churches developed—one in Western Europe and the other in the Byzantine empire.**

• The western church used Latin, while the eastern church used Greek.

> **The Byzantine empire slowly shrank and fell 800 years later.**

👆 **INTERACTIVE**

For extra help, review the 21st Century Skills Tutorial: **Draw Conclusions**.

Practice Vocabulary

Sentence Builder Finish the sentences below with a key term from this section. You may have to change the form of the words to complete the sentences.

Word Bank

Great Schism   Justinian's Code

missionary   icon

1. The western church and the eastern church became permanently split as a result of the

2. The western and eastern churches decorated church buildings with

3. People who try to convert others to a particular religion are called

4. In 529, Roman law was collected and organized into a few texts known as

Quick Activity Prepare a Statement

Every major event throughout history has had many causes and many effects. Select one of the four pictures below and write a brief statement that explains what it illustrates and its causes and effects.

▲ Humans use bronze to make tools and weapons.

▲ The Greeks develop city-states.

▲ The church in the Byzantine empire is closely connected to the government.

▲ Caesar Augustus becomes first Roman emperor.

Team Challenge! After everybody in your class posts their statements around the classroom, read them. Discuss with your classmates why they chose to write what they did. Do you have different ideas about the causes and effects your classmates chose? Discuss them.

Take Notes

Literacy Skills: Identify Cause and Effect Use what you have read to complete the charts. In the bottom boxes, enter three causes of each listed effect. The first one has been started for you.

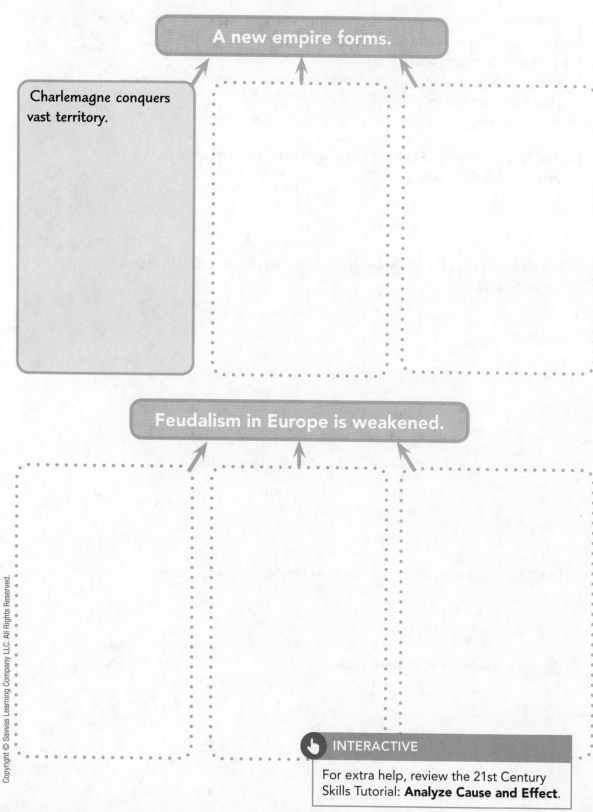

A new empire forms.

Charlemagne conquers vast territory.

Feudalism in Europe is weakened.

INTERACTIVE

For extra help, review the 21st Century Skills Tutorial: **Analyze Cause and Effect**.

Practice Vocabulary

Sentence Revision Revise each sentence so that the underlined vocabulary word is used logically. Be sure not to change the vocabulary word. The first one is done for you.

> 1. After the collapse of the western Roman empire, a system of marketplace reform called <u>feudalism</u> was established.
> *After the collapse of the western Roman empire, a system of government called <u>feudalism</u> was established.*

2. Most lords in medieval Europe were <u>serfs</u> who were required to live and work at a specific manor.

3. The disease called the <u>Black Death</u> contributed to a population increase in medieval Europe.

4. Under feudalism, <u>lords</u> kept land to themselves.

5. The heart of the medieval economy was the <u>manor</u>, which was a town center.

6. The <u>Crusades</u> began after Charlemagne's defeat of feudal lords.

7. <u>Vassals</u> were the most powerful lords.

Take Notes

Literacy Skills: Use Evidence Use what you have read to complete the table. In each column write details about the topic provided. The first one has been completed for you. Then, use the information you have gathered to draw a conclusion about the question provided.

Renaissance	Age of Discovery	Reformation
• Italians discovered teachings of ancient Greeks and Romans and learned new ideas.		
• Thinkers began to rely more on reason than on religious authority to understand the world.		
• Art became more realistic and scientists made great discoveries.		
• William Shakespeare wrote brilliant plays and poems.		

Conclusion: How did changes in Europe from the 1300s to the 1500s affect the way people understood the world and practiced religion?

INTERACTIVE

For extra help, review the 21st Century Skills Tutorial: **Identify Evidence**.

Practice Vocabulary

Matching Logic Using your knowledge of the underlined vocabulary terms, draw a line from each sentence in Column 1 to match it with the sentence in Column 2 to which it logically belongs.

Column 1	Column 2
1. The Protestant <u>Reformation</u> started as a movement to reform the Catholic Church.	This cultural flowering included new artistic methods as well as new ways of looking at the world.
2. French King Louis XIV embraced the idea of <u>absolute monarchy</u>.	Luther challenged the church's authority and stressed spirituality.
3. Italians rediscovered the learning of the ancient Greeks and Romans, spurring the <u>Renaissance</u>.	Missionary orders, such as the Jesuits, were formed.
4. People who believed in <u>humanism</u> sought a better life on earth.	Under this belief, the king had an unlimited right to rule.
5. During the <u>Counter-Reformation</u>, reformers founded religious groups with their own particular structure and purpose.	Thinkers began to focus on improving life on earth instead of just achieving salvation after death.

Take Notes

Literacy Skills: Classify and Categorize Use what you have read about the Enlightenment and revolutions in Europe to complete the tables. Categorize events based on the subject matter. The first one has been started for you.

Scientific Revolution	The Enlightenment
• Scientists started to use the scientific method to determine the truth. • Newton used the scientific method to prove the existence of gravity.	

Nationalism	Imperialism

🖐 **INTERACTIVE**

For extra help, review the 21st Century Skills Tutorial: **Categorize**.

Practice Vocabulary

Words in Context For each question below, write an answer that shows your understanding of the boldfaced key term.

1. What does a supporter of **nationalism** want?

2. What was the **Industrial Revolution**?

3. What happened during the **Enlightenment**?

4. What is **imperialism**?

5. How did the **French Revolution** begin?

Take Notes

Literacy Skills: Analyze Text Structure Use what you have read to complete the outlines. Fill in each outline to summarize the main ideas of the lesson. The outline has been started for you.

I. Why Were There Two World Wars?

   A. World War I

      1. Allied powers of France, Britain, Russia, and the United States fought and defeated Germany, Austria-Hungary, Bulgaria, and the Ottoman Empire; the defeated countries lost land and had to pay heavy reparations.

   B. Inflation and Depression

      1. Reparations led to hyperinflation in defeated countries, and the U.S. stock market crashed in 1929.

      2.

   C. Totalitarian Governments

      1.

      2.

      3.

   D. World War II

      1.

      2.

   E. Effects of World War II

      1.

      2.

      3.

 INTERACTIVE

For extra help, review the 21st Century Skills Tutorial: **Summarize**.

Practice Vocabulary

Vocabulary Quiz Show Some quiz shows ask a question and expect the contestant to give the answer. In other shows, the contestant is given an answer and must supply the question. If the blank is in the Question column, write the question that would result in the answer in the Answer column. If the question is supplied, write the answer.

Question	Answer
1. What is the economic and political partnership among member nations in Europe?	1.
2. What name was given to the mass murder of Jews and others by Nazis during World War II?	2.
3.	3. Cold War
4.	4. hyperinflation

Quick Activity Nazism and World War II Timeline

Write a timeline of events during the rise of Nazism and of the outbreak of World War II. Then, write one or two sentences about the effect on the people of Europe during the 1930s and 1940s.

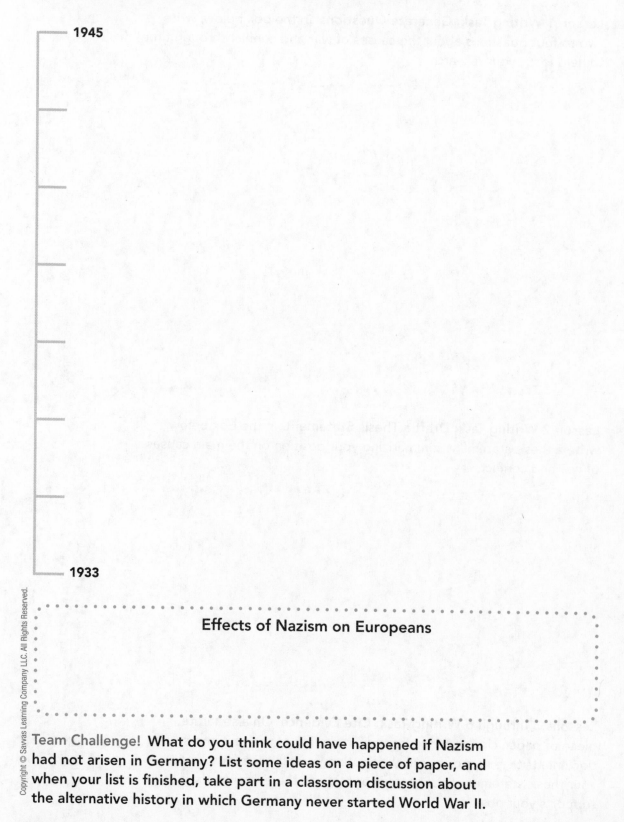

1945

1933

Effects of Nazism on Europeans

Team Challenge! What do you think could have happened if Nazism had not arisen in Germany? List some ideas on a piece of paper, and when your list is finished, take part in a classroom discussion about the alternative history in which Germany never started World War II.

Writing Workshop Explanatory Essay

As you read, build a response to this question: **What causes war and conflict?** The prompts below will help walk you through the process.

Lesson 1 Writing Task: Generate Questions In the box below, write two to four questions about the causes of war and conflict through time to help focus your research.

Lesson 2 Writing Task: Draft a Thesis Statement In the box below, write a thesis statement summarizing your position on the main causes of war and conflict.

Lessons 3 through 6 Writing Task: Cite Evidence On a separate piece of paper, cite evidence from your reading of the text and source documents that supports your thesis statement. If necessary, revise your thesis statement to align your position to the evidence that supports your position.

Lesson 7 Writing Task: Compare and Contrast List three wars or conflicts described in the lesson. In a few sentences, compare and contrast the causes of these wars or conflicts.

Lesson 7 Writing Task: Write an Introduction Write a paragraph that introduces your explanatory essay and includes your thesis statement.

Writing Task Using your thesis statement, the evidence you gathered, and your introduction, answer the following question in a five-paragraph explanatory essay: What causes war and conflict?

As you write, consider using the following cause-and-effect signal words to transition between points: *because, consequently, therefore, for this reason,* and *as a result.*

6 Europe Today Preview

Essential Question What makes a culture unique?

Before you begin this topic, think about the Essential Question by completing the following activities.

1. Briefly describe what culture means to you. How would you describe your culture, and what makes it different from other cultures?

2. Preview the topic by skimming lesson titles, headings, and graphics. Then place a check mark next to the factors that you predict the text says contribute to the existence of many unique cultures in Europe.

__language   __environment   __government

__religion   __ecotourism   __economy

__jazz   __migration   __geographic barriers

Map Skills

Using the Regional Atlas maps in your text, label the outline map with the places listed. Use a symbol to indicate urban areas with populations of more than 5 million, and create a key that explains your symbol. Then color in the land areas of Europe and the bodies of water.

Athens   Brussels   Bucharest   Budapest

Kiev   Lisbon   London   Madrid

Milan   Paris   Rhine-Ruhr   Warsaw

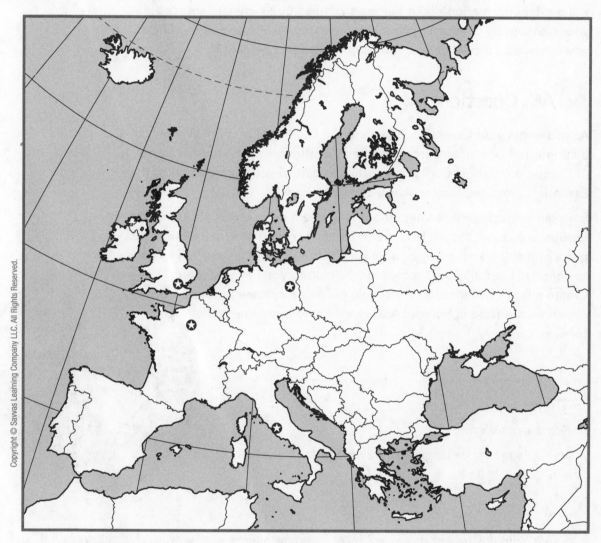

Create a Museum Exhibit

On this Quest, you have been asked by a museum curator to work with a team to create an exhibit about cultural diversity in modern Europe. You will gather information about Europe's cultures by examining sources in your text and by conducting your own research. At the end of the Quest, you will build a model of your museum exhibit and present it.

1 Ask Questions

As you begin your Quest, keep in mind the Guiding Question: **What distinguishes one culture from another?** Consider how the many cultures of Europe are distinct from one another as part of your exploration of the Essential Question: **What makes a culture unique?**

For your project, each team member will collect information about a different European culture that will be used to create part of a museum exhibit. To get a better understanding of how each culture is unique, you will need to consider the following themes for the culture you have selected. Create a list of questions that will help guide your research. Two questions are filled in for you. Add at least two more questions for each category.

Theme Language and Religion

Sample questions:

What language or languages are spoken by individuals in this culture?

What religious beliefs, if any, are most common among people from this culture?

Theme History and Settlement Patterns

Theme Family and Social Structure

Theme Arts, Beliefs, and Values

Theme My Additional Questions

 INTERACTIVE

For extra help with Step 1, review the 21st Century Skills Tutorial: **Ask Questions**.

2 Investigate

As you read about modern Europe, collect five connections from your text to help you answer the Guiding Question. Three connections are already chosen for you.

Connect to Language

Lesson 1 Why Does Europe Have So Many Languages?

Here's a connection! Read the section on Europe's languages. How did history contribute to the development of diverse languages in the past? What role did geographic barriers play?

How do you think modern transportation and communication methods have overcome geographic barriers and affected Europe's linguistic and cultural diversity?

Connect to Migration

Lesson 3 How Is the European Union Run?

Here is another connection! Read about the government of the European Union in the infographic. How do European Union policies affect the movement of people among member countries?

What effects does this movement have on cultural diversity?

Connect to Ethnic Identity

Lesson 4 Challenges Facing the European Union

Read about criticisms of the European Union. How has ethnic identity contributed to those criticisms?

How have economic conditions contributed to those criticisms?

It's Your Turn! **Find two more connections. Fill in the title of your connections, then answer the questions. Connections may be images, primary sources, maps, or text.**

Your Choice | Connect to

Location in text

What is the main idea of this connection?

What does it tell you about cultural diversity?

Your Choice | Connect to

Location in text

What is the main idea of this connection?

What does it tell you about cultural diversity?

3 Conduct Research

Form teams based on your teacher's instructions. As a team, decide which cultures to showcase and who will research each culture. The cultures your group chooses to research should include cultures of historically European ethnic groups and at least one example of a more recent immigrant culture.

You will do further research on the culture that you are responsible for. Use the Quest themes and connections to explore the culture that you selected. Find reliable primary and secondary sources about the culture. Record sources and key information in the table below. You may want to investigate additional themes to get a fuller understanding of the culture.

Theme	Source(s)	Notes
Language and Religion		
History and Settlement Patterns		
Family and Social Structure		
Beliefs and Values		
Arts		

INTERACTIVE

For extra help, review the 21st Century Skills Tutorials: **Evaluate Web Sites** and **Search for Information on the Internet**.

4 Create Your Exhibit

Now it's time to put together all of the information you have gathered and then plan and create your exhibit.

1. **Prepare to Create** Before you start to build your exhibit, organize your information and make a plan to help ensure that your exhibit is fact-based, logically organized, and eye-catching.

Culture:

Key information about the culture:

Sources to cite:

Possible visual or audio material to use:

2. **Create a Diagram** To plan your exhibit as a team, each member will make a diagram. The diagrams will be sketches or layouts that show how your exhibit will look. Begin by agreeing on a plan for the overall exhibit, and then independently develop a diagram for your own part. Your diagram should include a title, text, and at least one visual element. Visuals should have captions. Remember to plan your part with the complete exhibit in mind.

3. **Share** Share your diagram with the team. Discuss how all the diagrams work together and whether any changes are needed. Revise your diagram based on the team discussion.

4. **Build a Model** Once all team members have finalized plans, work together to create an exhibit model to present to the class.

5. **Present Your Exhibit** Present your exhibit model to the class. View the other teams' exhibits, and take notes on the information they shared.

Notes on other exhibits:

6. **Reflect** After viewing all the exhibit models, share your thoughts. What did you learn about cultural diversity in Europe from your team's research and the exhibit models presented by other teams? Which culture or cultures would you like to learn more about?

Reflections:

 INTERACTIVE

For extra help, review the 21st Century Skills Tutorials: **Give an Effective Presentation** and **Work in Teams**.

Take Notes

Literacy Skills: Identify Cause and Effect Use what you have read to complete the charts. In each space write details about the causes that have led to the identified effect. The first chart has been started for you.

Europe's Language Diversity

Indo-European Languages

Invaders brought Proto-Indo-European language to Central Europe about 5,000 years ago. Dialects spread across Europe and grew into distinct languages.

Influence of the Roman Empire

Impact of Geography

Europe's Religious Diversity

Judaism and Christianity

Islam and Protestantism

Recent Changes

INTERACTIVE

For extra help, review the 21st Century Skills Tutorial: **Analyze Cause and Effect**.

Practice Vocabulary

Words in Context For each question below, write an answer that shows your understanding of the boldfaced key term.

1. Explain how Proto-Indo-European **dialects** led to Europe's language diversity.

2. Why are most Basques **bilingual**?

Take Notes

Literacy Skills: Analyze Text Structure Use what you have read to create an outline of the main ideas of the lesson. As you create your outline, pay attention to headings, subheadings, and key terms that you can use to organize the information. The first section of the outline has been completed for you.

I. How Geographic Features Affect Where People Live
   A. Settlement Patterns
      1. Many early Europeans were farmers, and areas with rich farmland are still home to large populations.
      2. In western and northern Europe, people live near the coast where trade routes are easier to access and the climate is milder.
   B.

 INTERACTIVE

For extra help, review the 21st Century Skills Tutorial: **Identify Main Ideas and Details.**

Practice Vocabulary

Matching Logic Using your knowledge of the underlined vocabulary terms, draw a line from each sentence in Column 1 to match it to the sentence in Column 2 to which it logically belongs.

Column 1	Column 2
1. <u>Entrepreneurship</u> is strong in the German and British economies.	Free university education in her country helped Anja complete a degree in computer science.
2. Many Europeans are employed in <u>service sector</u> jobs.	María works for the public transportation system as a bus driver.
3. One way for a country to invest in <u>human capital</u> is by providing its people with quality education.	After Malik moved to England, he started his own international importing business.

Quick Activity Create a Living Population Density Map

The table below shows how the population of Europe would be distributed if the total population of Europe were 20. For example, 3 of the 20 would live in France.

Population of Europe in Terms of 20 People Total		
France: 3/20	Germany: 3/20	Italy: 3/20
Netherlands: 1/20	Poland: 2/20	Romania: 1/20
Spain: 2/20	Ukraine: 2/20	United Kingdom: 3/20

Calculate how many students in your class would live in each country based on these proportions. To do so, multiply the proportion for each country by the number of students in your class and round to the nearest whole number.

For example, if there are 27 students in your class, you would multiply 3/20 by 27 students. That gives a result of 4.05. After rounding, you know that 4 students would represent the population of France.

Record the proportions for your class in the table below.

Number of Students to Represent Countries' Populations		
France:	Germany:	Italy:
Netherlands:	Poland:	Romania:
Spain:	Ukraine:	United Kingdom:

Team Challenge! Create a living population density map using the proportions that you calculated. Work together to assign each student a country. Look at a map of Europe. Arrange yourselves around the classroom by country to form a giant map, with students assigned to each country standing together. Look around and discuss as a class: Where is population most concentrated? Which countries are missing? Why might those countries have lower populations? In which parts of Europe are people most concentrated?

Take Notes

Literacy Skills: Classify and Categorize Use what you have read to complete the chart. In each space write details about a type of government in Europe. The first one has been started for you. Provide examples of countries with each form of government.

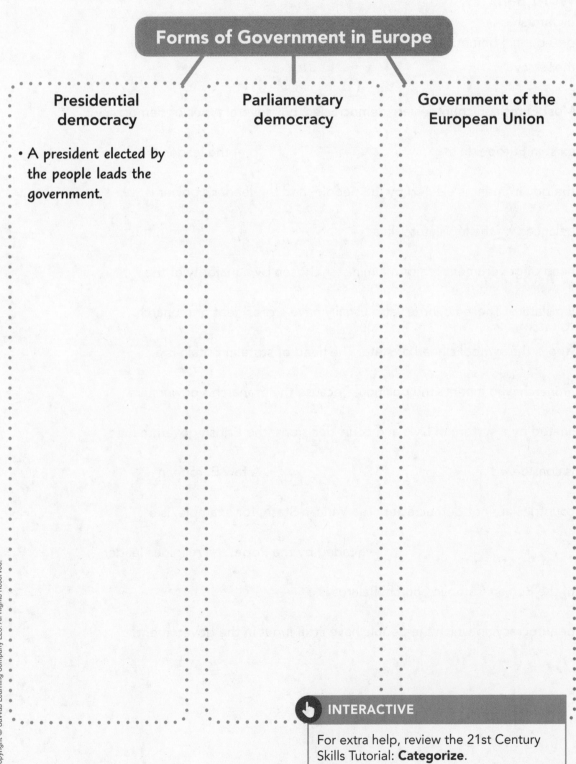

Forms of Government in Europe

Presidential democracy

• A president elected by the people leads the government.

Parliamentary democracy

Government of the European Union

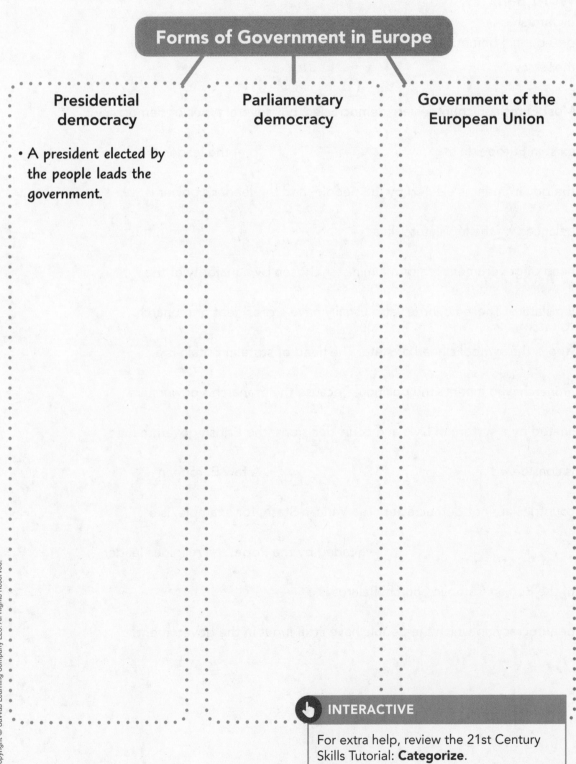

👆 **INTERACTIVE**

For extra help, review the 21st Century Skills Tutorial: **Categorize**.

Practice Vocabulary

Use a Word Bank Choose one word from the word bank to fill in each blank. When you have finished, you will have a short summary of important ideas from the section.

Word Bank

dictatorship   constitutional monarchy
presidential democracy   parliamentary democracy
theocracy

Most European countries are democracies, but several types of democracy exist in Europe. In a ...., the leader of the government is elected by the people, and the leader's power is balanced by the legislature. In a ...., the head of government is a prime minister chosen by a majority of the legislature. These countries also usually have a president or monarch who is the symbolic head of state. The head of state in Britain is a monarch who inherits the position. Because the monarch's powers are limited by a system of laws and court decisions, the British government is considered a .... A few European countries are not democracies. The Vatican State, for example, is a .... headed by the Pope, the religious leader of the Roman Catholic Church. Belarus is a ...., or autocracy, in which the people have little input in the government.

Take Notes

Literacy Skills: Draw Conclusions Use what you have read to complete the table. In each column write details about the challenges European countries are facing. The first details have been completed for you. Then use the notes you have taken to draw a conclusion related to the challenges facing Europe today.

Environmental challenges	Demographic challenges	Challenges facing the European Union
Air pollution	Aging population	Criticism of the EU
• harms crops, forests, and human health • causes acid rain that damages buildings, forests, and water supplies		
Nuclear fears	Economic effects	The euro crisis
Climate change		

Conclusion

INTERACTIVE

For extra help, review the 21st Century Skills Tutorial: **Draw Conclusions**.

Practice Vocabulary

Sentence Builder Finish the sentences below with a key term from this section. You may have to change the form of the words to complete the sentence.

Word Bank

climate change   demographic

greenhouse gases   Kyoto Protocol

1. Products of burning fossil fuels that scientists believe trap heat in Earth's atmosphere and make the planet warmer are called

   ....

2. The international agreement the European Union signed in 1997 pledging to reduce the emission of greenhouse gases is the

   ....

3. If data have to do with population change and the groups that make up the population of a place, those data are

   ....

4. A long-term increase in Earth's temperature is one example of global

   ....

Quick Activity Discussing Government Spending

An issue that every country faces is the question of government spending: How much should the government spend, and what should it spend its money on? How much spending is too much, and when is more spending justified? For countries in the European Union, there is another question: How will spending by the government of one member nation affect the other member nations?

Using what you have learned in this lesson, briefly write arguments for each of the positions provided in the table.

PRO: Government spending creates jobs and boosts the economy.	CON: Too much taxing and spending weakens the economy and is a burden.

Team Challenge! Form pairs and discuss the role of government spending and its effects on individual countries and on the European Union as a whole. Your teacher may assign you a side to take in this discussion. After discussing the issue with your partner, use the space below to record your opinion on the European Union's spending.

Writing Workshop Explanatory Essay

As you read, build a response to this question: **What is the impact of cultural diversity on Europe?** The prompts below will help walk you through the process.

Lesson 1 Writing Task: Develop a Clear Thesis Write a statement about the impact of cultural diversity in Europe. This sentence will be the thesis statement, or main idea, for the explanatory essay you will write at the end of the topic. Think about what you have read so far about Europe's diverse cultures and the effects of that diversity.

Lesson 2 Writing Task: Support Thesis with Details Record details from this lesson and the previous lesson that support your thesis. Continue recording details that support your thesis in the remaining lessons of the topic.

Lesson 3 Writing Task: Write an Introduction Write a paragraph introducing your thesis about the impact of cultural diversity in Europe and three main points that you will make to support your thesis based on what you have learned in this topic. You may need to revise your thesis as you learn more. This paragraph will be the introduction to your essay.

Lesson 4 Writing Task: Clarify Relationships with Transition Words Using transition words like *additionally* and *furthermore* strengthens writing by helping the reader see how ideas presented in a piece are connected. On a separate sheet of paper, create a rough outline for your essay, and write sentences that might connect parts of your outline using transition words.

Writing Task Using your thesis, introduction, and rough outline, answer the following question in a five-paragraph explanatory essay: What is the impact of cultural diversity on Europe? Incorporate some of the transition words you noted. For the conclusion, revisit your thesis and explain why the information in your essay is important. Re-read your essay carefully and edit it for proper spelling, grammar, and punctuation.

Essential Question What role should people have in their government?

Before you begin this topic, think about the Essential Question by completing the following activities.

1. Think about your community, state, or nation as a whole. List at least three ways that people participate in government. Then write a sentence about how you think people should participate in their government and explain why you think so.

Map Skills

Using the political and physical maps in the Regional Atlas in your text, label the outline map with the places listed.

Turkmenistan	Caspian Sea
Armenia	Georgia
Uzbekistan	Russia
Black Sea	Azerbaijan
Tajikistan	St. Petersburg
Kyrgyzstan	Kazakhstan
Moscow	Ural Mountains
Siberia	Arctic Ocean

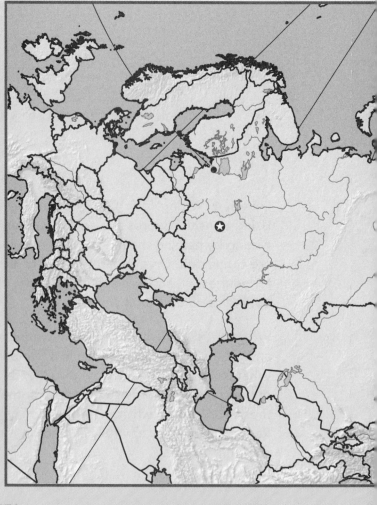

2. Preview the topic by skimming lesson titles, headings, and graphics. Then place a check mark next to the types of governments in Northern Eurasia that you predict will be discussed in the text. After you finish reading the topic, circle the predictions that were correct.

__parliamentary democracy

__dictatorship

__feudalism (serfdom)

__economic oligarchy

__communist regime

__theocracy

__direct democracy

__monarchy

__presidential democracy

__authoritarian regime

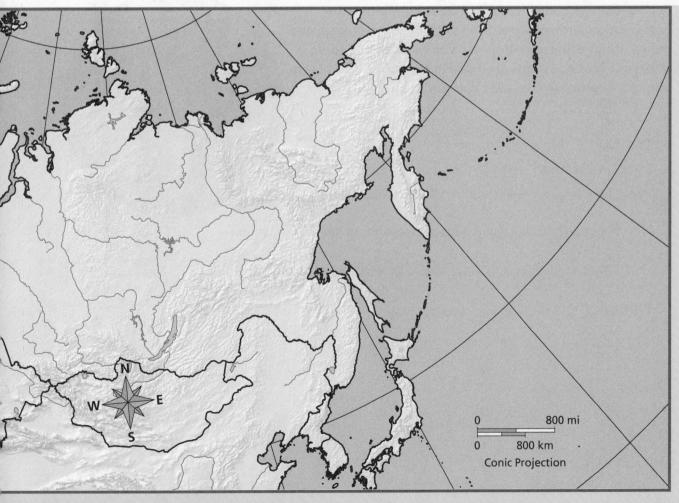

N
W E
S

0   800 mi

0   800 km

Conic Projection

Quest

Project-Based Learning Inquiry

Evaluating the Soviet Legacy

On this Quest, you will work with a team to put together a multimedia presentation explaining how Soviet economic decisions affected the region's natural environments and what challenges the Soviet system created for the region's economies. You will gather information about contemporary regional issues by examining sources in your text and by conducting your own research. At the end of the Quest, you will create a multimedia presentation to educate others.

1 Ask Questions

As you begin your Quest, keep in mind the Guiding Question: **How has the Soviet Union left a mark on the economies and environments of Northern Eurasia?** and the Essential Question: **What role should people have in their government?**

For your project, each team will create a multimedia presentation with a segment on each of the themes listed below. What other questions do you need to ask to do this task? Two questions are filled in for you. Add at least two more questions for each category.

Theme Government Systems

Sample questions:

What kind of government did the Soviet Union have?

How did governments change after the collapse of the Soviet Union?

Theme Agriculture and Industry

Theme Trade

Theme Environment and Health

Theme My Additional Questions

 INTERACTIVE

For extra help with Step 1, review the 21st Century Skills Tutorial: **Ask Questions**.

Quest CONNECTIONS

2 Investigate

As you read about Northern Eurasia, collect five connections from your text to help you answer the Guiding Question. Three connections are already chosen for you.

Connect to Command Economy

Lesson 2 Transition from Communism

Here's a connection! Examine the structure of the Soviet economy. What were the characteristics of the Soviet command economy?

How did those characteristics affect attempts to transition to a market economy after the breakup of the Soviet Union?

Connect to Market Economy

Lesson 4 Russia's Economy

Here's another connection! Take a closer look at the Russian economy today. What are some strengths?

How do you think it shows lingering effects of the Soviet system?

Connect to Challenges in the Region

Lesson 5 Political Challenges

What does this connection tell you about the challenges the Soviet Union left behind in Northern Eurasia?

How do these challenges affect the economies and the environment of the countries in the region?

It's Your Turn! Find two more connections. Fill in the title of your connections, then answer the questions. Connections may be images, primary sources, maps, or text.

Your Choice | Connect to

Location in text

What is the main idea of this connection?

What does it tell you about how the Soviet Union affected the environments and economies of the region?

Your Choice | Connect to

Location in text

What is the main idea of this connection?

What does it tell you about how the Soviet Union affected the environments and economies of the region?

③ Conduct Research

Form teams based on your teacher's instructions. Meet to decide who will create each segment of your presentation. Use the ideas in the connections to further explore the subject you have been assigned. Pick what you will report about, and find more sources about that subject.

Turn to the Country Databank at the end of your student text for data on the countries in the region. Look to sources like the CIA World Factbook or the World Bank's Databank for more current information and statistics. Take good notes so you can properly cite your sources. Brainstorm ways to express your findings by using various kinds of media.

Segment	Source	Notes
Government Systems		
Agriculture and Industry		
Trade		
Environment and Health		

INTERACTIVE

For extra help, review the 21st Century Skills Tutorials: **Analyze Media Content** and **Create Charts and Maps**.

Quest FINDINGS

4 Create Your Multimedia Presentation

Now it's time to put together all of the information you have gathered and create your multimedia presentation.

1. **Prepare to Write** Draw conclusions from the research you've collected and determine the main idea of your segment. Select one or two multimedia pieces that you will use to present your main idea. Below, write down your main idea and then record what two pieces you will use to support it, with a sentence explaining how each supports your main idea.

2. **Write a Draft** Write a short paragraph explaining your idea, then captions for each piece of media, summarizing the story that it tells. Clarify how it illustrates your theme in relation to the Guiding Question and the Essential Question. Include key facts, and cite sources.

3. **Plan and Revise with Your Team** Share your draft with your team. Discuss how you will combine your segments and the story they tell as a whole. Consider if anything needs to change to make the group presentation cohesive and concise. Plan how you will organize everyone's work. Remember, comparing and contrasting can help present opposing views in a cohesive way. Break down into pairs to examine each other's work in detail and to make changes discussed by the team.

4. **Share with Your Class** Follow teacher instructions for sharing your multimedia presentation. Include material from each team member. Present your project according to your team's organizational plan. Experience other presentations and note at least one impact each one had on you.

Notes on other presentations:

5. **Reflect** Think about your experience completing this topic's Quest. What did you learn about the effect of the Soviet Union on the environments and economies of the region? What questions do you still have? How will you answer them?

Reflections

 INTERACTIVE

For extra help, review the 21st Century Skills Tutorials: **Give an Effective Presentation** and **Work in Teams**.

Take Notes

Literacy Skills: Summarize Use what you have read to complete the table. List important events or facts that describe how Russia formed, grew, and declined. The first detail has been completed for you. Then write a few sentences summarizing Russia's cultural history before World War I.

How Did Russia Form?	How Did Russia Grow?	How Did Russia Decline?
• Drawn by trade, Vikings settled in what would later become Ukraine, adopted Slavic practices, and eventually ruled the region as the Kievan Rus.		

Summary

> **INTERACTIVE**
>
> For extra help, review the 21st Century Skills Tutorial: **Summarize**.

Practice Vocabulary

Use a Word Bank Choose one term from the word bank to fill in each blank. When you have finished, you will have a short summary of important ideas from the section.

Word Bank

serfs	clans	inflation
tsar	Silk Road	westernize

Early in Russia's history, the settlers formed .... ,

or family groups that lived and traded along the Black Sea. As people

moved around and traded more, the .... ,

a trade network linking China to the Mediterranean Sea, emerged.

Russian nobles gained great wealth through the labor of

.... , who worked the land

but had no rights. Peter the Great became the first great

.... , or ruler, from the Romanov dynasty. He

tried to .... Russia and bring new technology

and culture in from western Europe. By the 1800s, Russia still depended

on agriculture. The tsar emancipated, or freed, the serfs, hoping that

would help Russia become more modern. A lot of people remained very

poor, however. Russia's entry into World War I led to food shortages and

.... , and people could not afford what little

food was available.

Take Notes

Literacy Skills: Sequence Use what you have read to complete the timeline. Using specific dates, list the important events of the rise and fall of communism in Russia. Then attach the boxes to the correct place on the timeline.

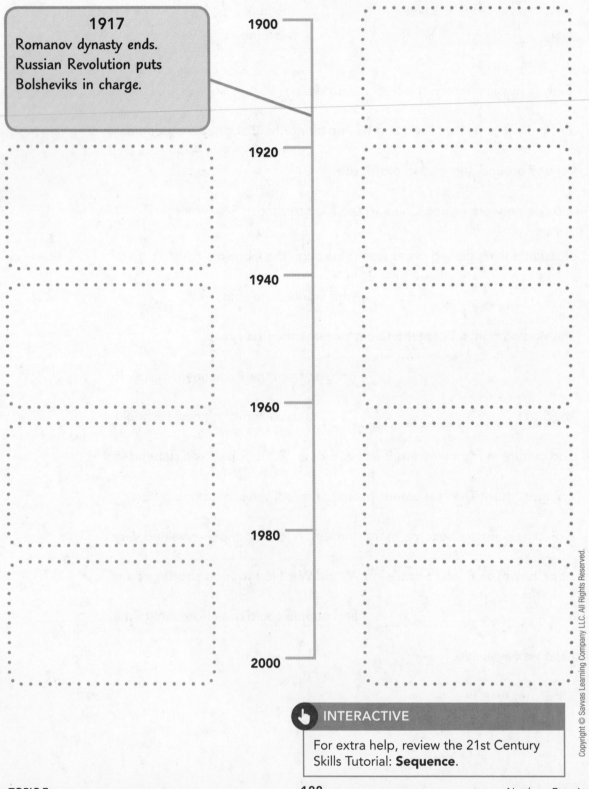

1917
Romanov dynasty ends.
Russian Revolution puts
Bolsheviks in charge.

1900

1920

1940

1960

1980

2000

👆 **INTERACTIVE**

For extra help, review the 21st Century
Skills Tutorial: **Sequence**.

Practice Vocabulary

Sentence Revision Revise each sentence so that the underlined vocabulary term is used logically. Be sure not to change the vocabulary term. The first one is done for you.

1. The <u>sanctions</u> placed on Russia were intended to permit it to do whatever it wanted to Crimea.
   The <u>sanctions</u> placed on Russia were intended to get Russia to follow international law.

2. The development of <u>communism</u> meant that property was owned by individuals.

3. When the Soviet Union fell, an <u>economic oligarchy</u> was formed by the workers.

4. The <u>Bolsheviks</u> called for the tsars to control Russia.

5. The <u>Cold War</u> began because the Soviet Union and the United States disagreed about how to fight World War II.

6. <u>Collectivization</u> transferred ownership of farms to individuals.

Quick Activity Debate!

Hold a debate with a partner in which one of you supports the ideals of communism and the other supports the policies of glasnost and perestroika. Use the table to help you prepare for your debate. Be sure to come up with arguments and counterarguments.

I support:	
Arguments:	
Counterarguments:	
Summarizing Argument:	

Team Challenge! Create a T-chart with your whole class. Each student supporting communism writes a reason on one side of the chart, and each student supporting the policies of perestroika and glasnost writes a reason on the other side of the chart. Discuss the reasons that might make both arguments valid. Have you changed your opinion?

Take Notes

Literacy Skills: Draw Conclusions Use what you have read to complete the charts. For each factual statement, write three conclusions that explain the significance of the fact. The first one has been completed for you.

More people live in cities than in Siberia, where there are more natural resources.

Living in Siberia is difficult because of the cold climate, and people generally prefer to live in places with a milder climate.

In the Caucasus, there are many ethnic groups that sometimes clash.

INTERACTIVE

For extra help, review the 21st Century Skills Tutorial: **Draw Conclusions**.

Practice Vocabulary

Word Map Study the word map for the word *permafrost*. Characteristics are words or phrases that relate to the word in the center of the word map. Non-characteristics are words and phrases not associated with the word. Use the blank word map to explore the meaning of the word *taiga*. Then make a word map of your own for the word *tundra*.

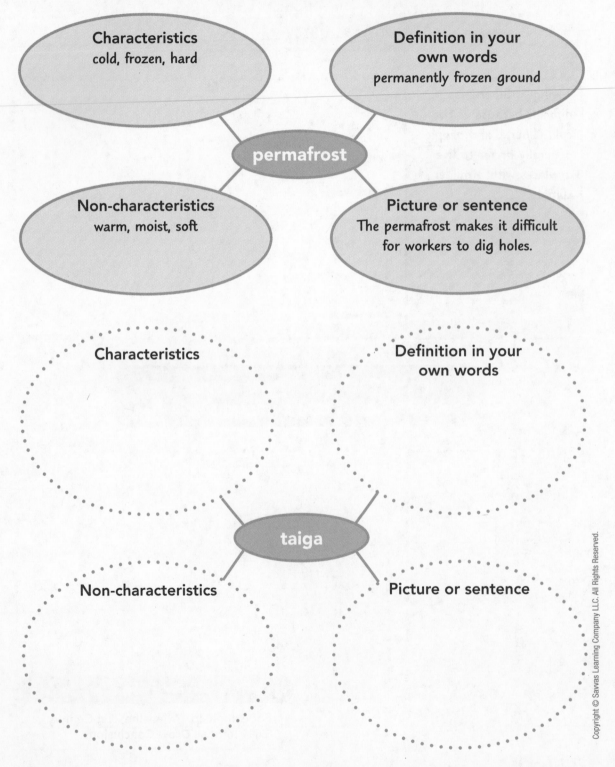

Characteristics
cold, frozen, hard

Definition in your own words
permanently frozen ground

permafrost

Non-characteristics
warm, moist, soft

Picture or sentence
The permafrost makes it difficult for workers to dig holes.

Characteristics

Definition in your own words

taiga

Non-characteristics

Picture or sentence

Take Notes

Literacy Skills: Use Evidence Use what you have read to complete the table. For each statement, provide three pieces of evidence from the text that explain why it is true. The first piece of evidence has been provided for you.

Statement	Evidence
Russia has an authoritarian government.	• Elections are not free and fair.
The effects of the Soviet Union continue to affect Russia's economy.	
Natural resources are critical parts of the economy for many former Soviet republics.	

INTERACTIVE

For extra help, review the 21st Century Skills Tutorial: **Identify Evidence**.

Practice Vocabulary

True or False? Decide whether each statement below is true or false. Circle T or F, and then explain your answer. Be sure to include the underlined vocabulary term in your explanation. The first one is done for you.

1. **T / F** <u>Industrial goods</u> are products or materials needed to make other products.
   True; An example of <u>industrial goods</u> is machinery made to produce paper.

2. **T / F** Having an <u>authoritarian government</u> means that the people govern themselves.

3. **T / F** Russia lacks <u>reserves</u> of gas, copper, and iron.

Quick Activity Categorize Governments

Since the fall of the Soviet Union, a few different types of government have emerged in the countries of Northern Eurasia. With a partner, use the table to categorize governments in Northern Eurasia. First identify the different forms of government. Then use the information in your text to list the countries that utilize each form of government.

Type of Government	Countries

Team Challenge! On a separate piece of paper, write what type of government you classified Russia as having and two reasons for doing so. Hang it up in your classroom. Did any teams classify it differently? Discuss the reasons there may be differences.

Take Notes

Literacy Skills: Identify Cause and Effect Use what you have read to complete the chart. For each cause provided for you, write an effect that might occur because of it. For each effect provided, write a major cause. The first one has been completed for you.

Cause	Effect
The Green Revolution involved heavy use of toxic fertilizers and pesticides that caused high levels of pollution across Russia.	Pesticides and fertilizers have killed fish and caused nearby communities to have high rates of cancer.
	The Aral Sea became too salty, the fish died, and the lake evaporated.
	Norilsk, a mining town, is one of the most polluted cities on the planet.
People who had been part of the Soviet government became leaders in new countries.	
Conflicts between ethnic groups persist across Central Asia.	
Corruption is a serious problem in many countries across Central Asia.	

INTERACTIVE

For extra help, review the 21st Century Skills Tutorial: **Analyze Cause and Effect**.

Practice Vocabulary

Sentence Builder Finish the sentences below with a key term from this section. You may have to change the form of the words to complete the sentences.

Word Bank

Green Revolution   militant   Aral Sea

1. A person who is aggressively active in a cause is called a

```
..............................................
.                                            .
.                                            .
.                                            .
..............................................
```

2. The term for the big increase in agriculture in the Soviet Union and other countries during the 1950s and 1960s is the

```
..............................................
.                                            .
.                                            .
.                                            .
..............................................
```

3. A body of water that has been destroyed by the effects of environmental damage is the

```
..............................................
.                                            .
.                                            .
.                                            .
..............................................
```

Writing Workshop Argument

As you read, build a response to this question: **What role should citizens play in their government?** The prompts below will help walk you through the process of turning your ideas into an essay presenting an argument.

Lesson 1 Writing Task: Introduce Claims Think about how Russia's history was affected by its citizens' role, or lack of a role, in government. Then, write a broad statement about what you think the role of citizens in government should be. You will develop this claim into a thesis statement for the argument you will write at the end of the topic.

Lessons 2 and 3 Writing Task: Support Claims Add support for the claim in your thesis from all of the lessons in this topic to the table below. For instance, how does the Soviet Union and its breakup, discussed in Lesson 2, support an argument for a strong or limited role for citizens in government? Does the cultural diversity of Northern Eurasia, discussed in Lesson 3, support a strong role for citizens? Adjust your thesis if the evidence you find pulls you in another direction.

Support from Lessons 1 and 2	
Support from Lesson 3	
Support from Lessons 4 and 5	

Lessons 4 and 5 Writing Task: Write an Introduction and Organize Your Essay Write an introductory paragraph in the first row of the table below that introduces your thesis and lays out what points you will make to support your argument. Then make a brief outline of your essay, including the order of your supporting points and addressing opposing claims. End with your strongest point.

Introduction	
Outline	

Writing Task: Using your introduction and outline above, write an argument that describes the role you believe citizens should have in their government. Support your claims with evidence from the text. Be sure to address opposing claims, and use transitional words or phrases to clarify the relationships between your ideas.

Acknowledgments

Photography

004 Igor Mojzes/Alamy Stock Photo; **006** Rick Dalton–Ag/Alamy Stock Photo; **007** Kristoffer Tripplaar/Alamy Stock Photo; **009** Dmitry Kalinovsky/Shutterstock; **010** Michael Warren/iStock/Getty Images; **036** Michel Loiselle/Alamy Stock Photo; **038** Todd Taulman/Shutterstock; **039** Image BROKER/Alamy Stock Photo; **040** Orhan Cam/Shutterstock; **041** Art Babych/Shutterstock; **044** IanDagnall Computing/Alamy Stock Photo; **067** J Gerard Sidaner/Science Source/Getty Images; **068** EDU Vision/Alamy Stock Photo; **069** Jose Goitia/Gamma-Rapho/Getty Images; **071** Omar Torres/AFP/Getty Images; **096** Raul Arboleda/Stringer/AFP/Getty Images; **098** EPA european pressphoto agency b.v./Alamy Stock Photo; **100** Deco/Alamy Stock Photo; **101** Pulsar Images/Alamy Stock Photo; **103** Frilet Patrick/Hemis/Alamy Stock Photo; **104** Yohei Osada/Nippon News/Aflo Sport/Alamy; **124** Derrick E. Witty/National Trust Photo Library/Art Resource, NY; **126** Marek Druszcz/AFB/Getty Images; **127** Lebrecht Music and Arts Photo Library/Alamy Stock Photo; **128** Lambros Kazan/Shutterstock; **129** World History Archive/Alamy Stock Photo; **142T** Zev Radovan/BibleLandPictures.com/Alamy Stock Photo; **142C** Lambros Kazan/Shutterstock; **142BL** Hercules Milas/Alamy Stock Photo; **142BR** Evannovostro/Shutterstock; **154** Eye35/Alamy Stock Photo; **156** VLIET/iStock Unreleased/Getty Images; **157** Ian G Dagnall/Alamy Stock Photo; **158** Eduardo Gonzalez Diaz/Alamy Stock Photo; **159** Alex Segre/Alamy Stock Photo; **162** Michael Brooks/Alamy Stock Photo; **178** SVF2/Universal Images Group/Getty Images; **180** Everett Collection Historical/Alamy Stock Photo; **181** ITAR-TASS Photo Agency/Alamy Stock Photo; **184** Shamil Zhumatov/Reuters/Alamy Stock Photo